TECHNICAL REPORT

T0166247

A Methodology for Implementing the Department of Defense's Current In-Sourcing Policy

Jessie Riposo, Irv Blickstein, Stephanie Young, Geoffrey McGovern, Brian McInnis

Prepared for the United States Navy

Approved for public release; distribution unlimited

RAND NATIONAL DEFENSE RESEARCH INSTITUTE

The research described in this report was prepared for the United States Navy. The research was conducted within the RAND National Defense Research Institute, a federally funded research and development center sponsored by the Office of the Secretary of Defense, the Joint Staff, the Unified Combatant Commands, the Navy, the Marine Corps, the defense agencies, and the defense Intelligence Community under Contract W74V8H-06-C-0002.

Library of Congress Cataloging-in-Publication Data

A methodology for implementing the Department of Defense's current in-sourcing policy / Jessie Riposo ... [et al.].
    p. cm.
  Includes bibliographical references.
  ISBN 978-0-8330-5311-4 (pbk. : alk. paper)
 1. United States. Dept. of Defense—Procurement. 2. United States. Dept. of Defense—Rules and practice.
 3. Contracting out—United States—Evaluation. I. Riposo, Jessie. II. Title.

   UC263.M445 2011
   355.6'21—dc23

                                                                            2011025827

The RAND Corporation is a nonprofit institution that helps improve policy and decisionmaking through research and analysis. RAND's publications do not necessarily reflect the opinions of its research clients and sponsors.

RAND® is a registered trademark.

Published 2011 by the RAND Corporation
1776 Main Street, P.O. Box 2138, Santa Monica, CA 90407-2138
1200 South Hayes Street, Arlington, VA 22202-5050
4570 Fifth Avenue, Suite 600, Pittsburgh, PA 15213-2665
RAND URL: http://www.rand.org/
To order RAND documents or to obtain additional information, contact
Distribution Services: Telephone: (310) 451-7002;
Fax: (310) 451-6915; Email: order@rand.org

Preface

For decades, the Department of Defense (DoD) has worked to identify the appropriate balance of contractors and government employees. In the 1970s, it began efforts to use contractors who were more cost-effective. These efforts continued throughout the 1980s and 1990s, but skepticism grew as cost savings became difficult to track and as questions regarding the appropriate functions of contractors arose. In his fiscal year 2010 budget message, Secretary of Defense Robert M. Gates called for growing the civilian workforce by replacing contractors with DoD civilian personnel.[1] In the 2010 *Quadrennial Defense Review*, DoD's ongoing in-sourcing efforts were described as a means of establishing "a balanced total workforce . . . that more appropriately aligns public and private sector functions."[2] This report aims to present a methodology for implementing current DoD guidance on in-sourcing. It also presents a review of both the history of DoD sourcing and current law and policy pertaining to that sourcing.

This research was sponsored by the Assistant Deputy to the Chief of Naval Operations for Integration and Resources and conducted within the Forces and Resources Policy Center of the RAND National Defense Research Institute, a federally funded research and development center sponsored by the Office of the Secretary of Defense, the Joint Staff, the Unified Combatant Commands, the Navy, the Marine Corps, the defense agencies, and the defense Intelligence Community. It should be of interest to persons concerned with workforce planning in DoD, especially those who are responsible for the development of policy.

For more information on this research, contact the principal authors of this report, Jessie Riposo (Jessie_Riposo@rand.org) and Irv Blickstein (Irv_Blickstein@rand.org). For more information on the RAND Forces and Resources Policy Center, see http://www.rand.org/nsrd/ndri/centers/frp.html or contact the director (contact information is provided on the web page).

---

[1] U.S. Department of Defense, "Defense Budget Recommendation Statement," as prepared for delivery by Secretary of Defense Robert M. Gates, Arlington, Va., Statement of Robert M. Gates, "Defense Budget Recommendation Statement," April 6, 2009.

[2] U.S. Department of Defense, *Quadrennial Defense Review Report*, February 2010a, p. 56.

Contents

Figures and Table

Figures

Table

Summary

Identifying the appropriate balance between contractor and government staff is not a new challenge for the government. However, recent direction from Department of Defense (DoD) leadership has led to increased awareness of the importance of finding this balance. In his fiscal year (FY) 2010 budget message, Secretary of Defense Robert M. Gates called for growing the civilian workforce by replacing contractors with DoD civilian personnel.[3] In the 2010 *Quadrennial Defense Review*, DoD's ongoing in-sourcing efforts were described as a means of establishing "a balanced total workforce . . . that more appropriately aligns public and private sector functions."[4] In August 2010, Secretary Gates called for a halt to the creation of Office of the Secretary of Defense (OSD) positions to replace contractor staff after FY 2010.[5] Although some may have believed that in-sourcing efforts were to be ceased, additional guidance clarified that in-sourcing efforts would continue. A three-year hiring freeze was implemented at OSD, the combatant commands, and the defense agencies, but in-sourcing was allowed to continue in the military departments.[6]

The research conducted by the RAND Corporation reviewed the recent history of out-sourcing and in-sourcing and assessed the current laws and policies pertaining to in-sourcing in order to develop a framework and methodology for applying the current laws and policies to arrive at a decision about in-sourcing. This research was initially undertaken as part of an in-sourcing assessment for a U.S. Navy organization. However, during research, we discovered a gap in the current in-sourcing guidance, and individuals tasked with in-sourcing still appeared to be struggling to interpret and implement this guidance. This report interprets commonly used terms, such as *inherently governmental*, and describes an approach to implementing current in-sourcing guidance.

A History of Government Sourcing

Until the mid-1990s, outsourcing predominated at DoD. For decades, the department embraced expanding roles for private contractors as a means of achieving efficiencies and accomplishing missions. Federal policy reflected the belief that government outputs could be improved by letting the private sector perform an increasing array of functions. In the three decades prior to

---

[3] U.S. Department of Defense, 2009.

[4] U.S. Department of Defense, 2010a, p. 56.

[5] Office of the Assistant Secretary of Defense (Public Affairs), "Sec. Gates Announces Efficiencies Initiatives," News, Release No. 706-10, August 9, 2010.

[6] Robert Brodsky, "Union Blasts Scaled Down Defense In-Sourcing Plan," September 30, 2010c.

the mid-1990s, laws and policies had been changed to reflect the preference that contractors increasingly perform activities for DoD. This contributed to a decline in the civilian workforce. Between 1989 and 2002, the DoD civilian workforce declined by about 38 percent, dropping from 1,075,437 civilians to 670,166.[7]

Although there have always been opponents of outsourcing, increased interest in reevaluating the workforce mix began to emerge in early 2000. A Defense Science Board task force reported that the "[r]apid downsizing during the last ten years" had been a catalyst for rethinking the balance between components of the "total force"—contractors, civilian personnel, and military personnel.[8] In-sourcing momentum increased with support from members of Congress, and significant in-sourcing initiatives began when the Obama administration took office. However, this renewed focus on in-sourcing required clarification of existing workforce planning guidance and policy and of the in-sourcing process.

Unfortunately, producing definitive guidance that can be used to objectively determine the appropriate balance between contractor and government staff has proven quite challenging. Most policy and guidance has reflected the preferences of the presidential administration. Although there are definitions of the criteria to be used to determine whether a function should be performed by a contractor or government civilian, these definitions vary, and the in-sourcing analyst must exercise judgment in interpreting them.

Current Policy

Law and policy restrict the use of contractors for functions that should be the exclusive or near-exclusive responsibility of government employees. These restrictions identify the nature of work that contractors may not perform or that they may perform only under specific circumstances. There are four major restrictions on the use of contractors:

- a prohibition on contractor performance of inherently governmental (IG) functions
- special rules about the use of contractors to perform functions closely associated with IG functions
- a prohibition on the use of personal-services contracts
- DoD-specific exemptions from private-sector performance of specific commercial functions identified in law and policy.

As of this writing, there is also a proposed new restriction on contractor performance of critical functions.

The underlying concept behind each of these restrictions is straightforward: Only government employees should wield the authority of the government. However, consistent implementation is challenged by the level of interpretation required to assess whether various functions meet these criteria.

---

[7] U.S. General Accounting Office, *DoD Personnel: DoD Actions Needed to Strengthen Civilian Human Capital Strategic Planning and Integration with Military Personnel and Sourcing Decisions*, Washington, D.C., GAO-03-475, 2003.

[8] Defense Science Board Task Force, *Human Resources Strategy*, Washington, D.C.: Department of Defense, 2000.

Our Approach

We reviewed the history of sourcing as well as current policy and guidance pertaining to in-sourcing. This material served as the basis for our development of an analytical approach to help government departments implement in-sourcing plans that comply with the most-current policy and the most-current federal restrictions on the use of contractors. Our approach is based on a 2009 Deputy Secretary of Defense memorandum that provides a decision-tree method and a defined set of criteria for in-sourcing (see Figure S.1).

However, to apply this approach to a population of contractors, we had to select and interpret definitions of the assessment criteria provided in existing guidance and instructions. In addition, to assess whether defined criteria were met, we had to construct and answer specific questions, such as "How is *inherently governmental* defined, and what types of questions should be asked in order to determine whether a function is inherently governmental?" Although this exercise may appear to be rather simple at first glance, developing a definition and set of questions that result in consistent assessment is actually quite challenging. The questionnaire we developed for civilian leadership to use in determining whether selected criteria for in-sourcing are met is reproduced in Appendix C.

Figure S.1
The In-Sourcing Assessment Process

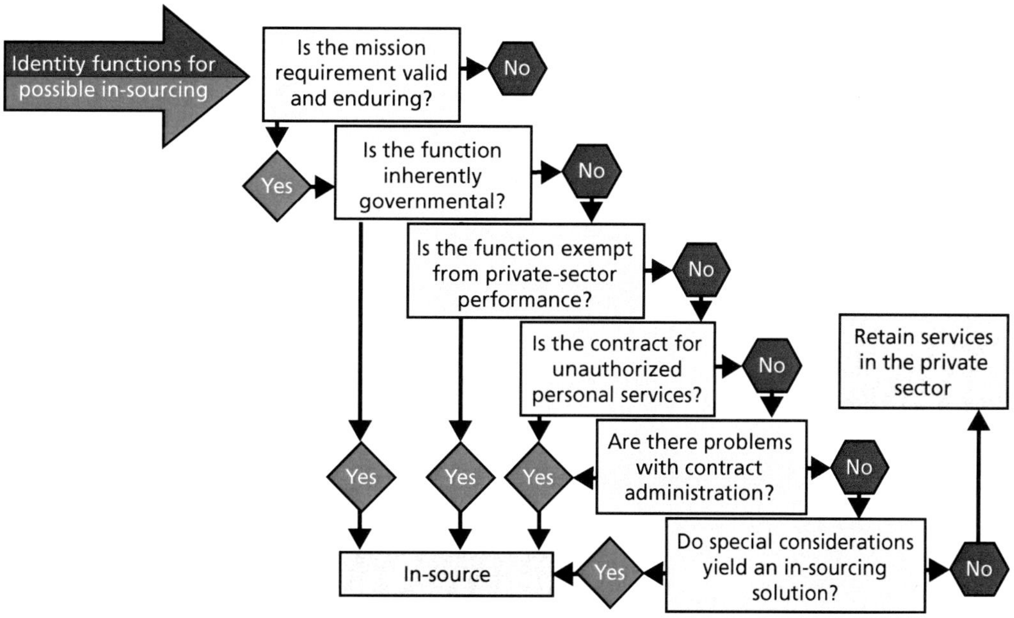

SOURCE: Based on Deputy Secretary of Defense, "In-Sourcing Contracted Services—Implementation Guidance," memorandum, May 28, 2009.
RAND *TR944-S.1*

Acknowledgments

We thank James McCarthy and Howard Fireman for sponsoring this work. We thank John Schank and Edward Keating for their thoughtful reviews of the draft manuscript.

Abbreviations

2009 NDAA	Duncan Hunter National Defense Authorization Act for Fiscal Year 2009
AAP	Acquisition Advisory Panel
CAAS	contracts for advisory and assistance services
CAWIG	closely associated with inherently governmental
CFR	Code of Federal Regulations
COR	Contracting Officer Representative
DFAR	Defense Federal Acquisition Regulation
DoD	Department of Defense
DoDI	DoD Instruction
FAIR Act	Federal Activities Inventory Reform Act
FAR	Federal Acquisition Regulation
FY	fiscal year
GAO	Government Accountability Office
IG	inherently governmental
NDAA	National Defense Authorization Act
OFPP	Office of Federal Procurement Policy
OMB	Office of Management and Budget
OSD	Office of the Secretary of Defense
PMA	President's Management Agenda
U.S.C.	United States Code
USD/P&R	Under Secretary of Defense for Personnel and Readiness

CHAPTER ONE

Introduction

Background

Contractors have been an important element of the Department of Defense's (DoD's) total-force mix since the 1970s. When stepping in to serve functions that are not inherently governmental (IG), contractors can complement civilian DoD personnel and provide workforce managers with increased flexibility, special expertise, and potential efficiencies. However, relative to the role played by government employees, the scope of contractor involvement has grown since the 1980s. The post–Cold War drawdown reduced the DoD workforce significantly, and contractors increasingly stepped in to fill functional gaps. DoD's increased reliance on contractors also reflected a broader perception in the U.S. government of the inefficiency of government institutions. The assumption that outsourcing to the private sector would yield cost savings may have motivated many decisions to contract for functions once conducted by civilian personnel. At the same time, there were questions about whether the decision to employ contractors was having the desired outcomes or was aligned with appropriate definitions of the IG concept.

In January 2010, when we began research, the Obama administration had made it a top priority to evaluate whether the current balance of in-sourcing and outsourcing was optimal or whether it would in fact be appropriate to decrease DoD's reliance on contractors. DoD leadership signaled the intent to reevaluate outsourcing decisions made in the past to ensure that the department strikes the appropriate balance between government and contract personnel. In his fiscal year (FY) 2010 budget message, Secretary of Defense Robert M. Gates called for growing the civilian workforce by replacing contractors with DoD civilian personnel,[1] and he reaffirmed that goal in the department's FY 2011 budget.[2] This posture was also reflected in the 2010 *Quadrennial Defense Review*, which described the department's ongoing in-sourcing efforts as a means of establishing "a balanced total workforce . . . that more appropriately aligns public and private sector functions."[3]

However, a June 2010 memorandum from Under Secretary of Defense Ashton B. Carter requested $28.3 billion in efficiencies between 2012 and 2015 from each of the three military

---

[1] U.S. Department of Defense, "Defense Budget Recommendation Statement," as prepared for delivery by Secretary of Defense Robert M. Gates, Arlington, Va., April 6, 2009.

[2] Robert M. Gates, "Statement on the Budget to the Senate Armed Services Committee," Washington, D.C., February 2, 2010.

[3] U.S. Department of Defense, *Quadrennial Defense Review Report*, Washington, D.C., February 2010a, p. 56.

departments.[4] A task force created to identify where to find these efficiencies is said to have recommended the reduction of 111,000 civilian employees across the department and a hiring freeze in the Office of the Secretary of Defense (OSD).[5] On August 9, 2010, Secretary Gates announced that the number of contractors performing IG functions would be reduced and that no more full-time OSD positions would be created after FY 2010 to replace contractors, except in the case of critical needs.[6]

It is unclear how these "critical needs" will be assessed, but there will clearly be an effect on all current in-sourcing efforts, including those discussed in this report. Despite the existence of implementation challenges, this report identifies important considerations for in-sourcing. The observations made should be of interest to anyone who is involved in workforce planning for DoD, especially those involved in policy development.

Research Objectives and Approach

Our research aims were to review the history of sourcing, describe current law and policy pertaining to in-sourcing, and develop an analytical approach for determining in-sourcing needs.

We performed an extensive literature review to support our research objectives. The analytical approach developed to determine the appropriateness of the conversion of positions from contractor to civilian was derived from a May 2009 memorandum entitled "In-Sourcing for Contracted Services—Implementation Guidance."[7] This document presented a decision-tree approach and criteria for determining whether a function should be in-sourced. However, implementing its guidance requires adopting definitions of the criteria, interpreting those definitions, and developing a questionnaire to assess whether positions meet the criteria. In practice, we recommend that interviews with civilian leadership and the contractors performing the work be conducted. In addition, the analyst should spend time observing the work environment.

Organization of This Report

Chapter Two chronicles the recent history of sourcing, showing that the majority of sourcing since the 1990s has taken the form of outsourcing. Chapter Three provides an overview of the law and policy pertaining to in-sourcing. Chapter Four presents an analytical approach that can be used to perform in-sourcing assessments. Chapter Five summarizes our findings.

---

[4]   Under Secretary of Defense, "Better Buying Power: Mandate for Restoring Affordability and Productivity in Defense Spending," Washington, D.C., June 28, 2010. Reductions occur in the following accounts: operations and maintenance, personnel, military construction, and revolving and management funds.

[5]   John T. Bennet, "Panel: DoD Should Cut 111,000 DoD Civilian Jobs," *Federal Times*, July 26, 2010, p. 1.

[6]   Office of the Assistant Secretary of Defense (Public Affairs), "Sec. Gates Announces Efficiencies Initiatives," News Release No. 706-10, August 9, 2010.

[7]   Deputy Secretary of Defense, "In-Sourcing for Contracted Services—Implementation Guidance," memorandum, May 28, 2009.

A Look at the History of Sourcing

Questions about the appropriate role of the private sector in the performance of governmental work have a long history.[1] This chapter, however, examines only DoD's recent experience in outsourcing and in-sourcing, beginning with the 1990s. This decade is a useful starting point because it coincided with substantial post–Cold War personnel reductions, which made growing the contractor workforce an attractive option. In addition, President Bill Clinton took office during that decade, shortly after the Office of Management and Budget (OMB) promulgated its influential definition of *inherently governmental functions*, and the term *inherently governmental* has become the basis for much of the discussion about sourcing.

The polices adopted by President Clinton and President George W. Bush expanded the opportunities for the private sector in government. The burden then fell on advocates of the civil service to justify why positions should *not* be outsourced rather why they should . However, there was pushback by the mid-2000s, when governmental reviews began calling into question the transparency, accountability, and cost-effectiveness of prior outsourcing decisions. In a move indicative of such concerns, Congress introduced in-sourcing language in the National Defense Authorization Act for Fiscal Year 2006.[2]

The incoming Obama administration supported in-sourcing initiatives, giving the issue new urgency and prominence. In general, the administration's language has characterized these initiatives as an effort to reverse outsourcing trends that had gone too far rather than as a fundamental reprioritization of the public sector. However, pushing the pendulum toward in-sourcing has proven significantly more difficult than pushing it toward outsourcing. The ease and speed of hiring contractors makes outsourcing appealing, and in-sourcing efforts (i.e., working within existent civil-service laws to hire contractors as government employees) are considerably more difficult and therefore less appealing. The first year and a half of the Obama administration witnessed a flurry of in-sourcing activity, but, by summer 2010, the issue had lost momentum, leaving the future of in-sourcing initiatives unclear.

Outsourcing

For decades, DoD has embraced expanding roles for private contractors as a means of achieving efficiencies and accomplishing missions. Since at least the Reagan administration, federal

---

[1] For a concise history of the topic see, John R. Luckey, Valerie Bailey Grasso, and Kate M. Manuel, *Inherently Governmental Functions and Department of Defense Operations: Background, Issues, and Options for Congress*, Washington, D.C.: Congressional Research Service, 2009.

[2] Public Law 109-163, National Defense Authorization Act for Fiscal Year 2006, January 6, 2006.

policy has reflected a sense that government outputs can be improved by letting the private sector perform an increasing array of functions. In addition, outsourcing has been motivated by the practical virtues of using contractors, such as flexibility in hiring and firing and access to specialized expertise. The outsourcing trend has not been a particularly Republican or Democratic initiative. Rather, both parties have enthusiastically promulgated outsourcing initiatives since the 1990s.[3]

In theory, outsourcing efforts must be balanced against competing government priorities. The full legal and regulatory framework governing such decisions is described in Chapter Three, but we wish to note that the rationale for limitations—and for changes in those limitation over time—is an important part of the history. The most widely discussed limitation on outsourcing concerns the nature of the function being performed. For example, in 2006, DoD Instruction (DoDI) 1100.22, *Policy and Procedures for Determining Workforce Mix*, identified certain functions as explicitly prohibited (i.e., exempt) from being performed by contractors, such as functions associated with civilian and military career development and those involving operational risk.[4] Most designations of out-of-bounds functions are less clear, however. For example, IG functions are off-limits to contractors. The term *inherently governmental* dates to 1966, when the first Circular No. A-76 was issued. This circular stated that "certain functions are inherently governmental in nature, being so intimately related to the public interest as to mandate performance only by federal employees."[5] Interpreting the slippery IG standard, as well as an additional class of functions known as *closely associated with inherently governmental* (CAWIG), has received by far the most attention in the recent history of sourcing. In 2010, OMB recommended an additional function-based limitation on the use of contractors. It recommended that certain functions be deemed "critical" by virtue of their importance to an agency's mission. At least a portion of such vital functions must be reserved for governmental performance to ensure that the agency has "sufficient internal capability to effectively perform and maintain control of its mission and operations."[6] These function-based limitations reflect a sense that, either in the "public interest" or in the interest of allowing the agency to control its own operations, certain functions must be off-limits to contractors.

Function-based rationales are not the only class of limitations on the role of contractors in the private sector. Since 1943, the government has promulgated a ban on personal-services contracts.[7] Such a contract is one that, either "by its express terms or as administered, makes the contractor personnel appear to be government employees."[8] Standards for identifying personal-services contracts were clarified in 1967 and incorporated into the Federal Acquisition Regulation (FAR) shortly thereafter. Unlike function-based limitations, the limitation on

---

[3] "Outsourcing is a decision by the government to purchase goods and services from sources outside of the affected government agency" (Valerie Bailey Grasso, *Defense Outsourcing: The OMB Circular A-76 Policy*, Washington, D.C.: Congressional Research Service, 2005).

[4] U.S. Department of Defense, *Policy and Procedures for Determining Workforce Mix*, DoDI 1100.22, April 12, 2010b.

[5] OMB, quoted in Acquisition Advisory Panel, *Report of the Acquisition Advisory Panel to the Office of Federal Procurement Policy and the United States Congress*, 2007.

[6] Office of Management and Budget, Office of Federal Procurement Policy, "Work Reserved for Performance by Federal Government Employees," web page, undated.

[7] R. E. Korroch, *Rethinking Government Contracts for Personal Services*, Washington, D.C.: George Washington University, 1997.

[8] Acquisition Advisory Panel, 2007.

personal-services contracts is justified by an appeal to the importance of remaining faithful to laws governing employment in the civil service. The limitation is not directly justified by the need to preserve either the public interest or an agency's ability to control its own operations; rather, it is explained by the need to preserve the civil service by maintaining regulations on hiring and firing and on personnel ceilings. Although the personal-services prohibition has received considerably less attention than function-based prohibitions, it may affect a significant portion of the contract workforce.

The Clinton Administration, 1993–2001

Since the end of the Cold War, the DoD civilian workforce has shrunk considerably. Between 1989 and 2002, the DoD civilian workforce decreased by about 38 percent, dropping from 1,075,437 civilians to 670,166.[9] In 2000, a Defense Science Board task force reported that the "[r]apid downsizing during the last ten years" had been a catalyst for rethinking the balance between components of the "total force"—contractors, civilian personnel, and military personnel.[10]

To define appropriate relationships for contractors and civilians, the government needed to draw clear boundaries between the two. In 1992, the Office of Federal Procurement Policy (OFPP) published a policy letter that defined IG functions, thereby drawing a line between those functions that had to be performed by the government and those that could be contracted out. The policy letter was written to prevent "an unacceptable transfer of official responsibility to Government contractors."[11] It affirmed that contractors, when used appropriately, can provide special knowledge, cost-effective services, and temporary support, but it asserted that some functions were out-of-bounds. It defined an IG function as one that "is so intimately related to the public interest as to mandate performance by Government employees."[12] Such functions include "activities that require either the exercise of discretion in applying Government authority or the making of value judgments in making decisions for the Government."[13] Despite efforts to codify IG functions, the reality is that most functions fall somewhere between extremes, meaning that there is significant space for interpretation in what constitutes an IG activity.[14]

The incoming Clinton administration prioritized new evaluations of the role of the private sector in government operations. In March 1993, President Clinton announced a six-month review of the federal government.[15] The President explained the National Performance Review (known since 1998 as the National Partnership for Re-Inventing Government) thus: "Our goal is to make the entire federal government both less expensive and more efficient. . . . We intend

---

[9] U.S. General Accounting Office, *DoD Personnel: DoD Actions Needed to Strengthen Civilian Human Capital Strategic Planning and Integration with Military Personnel and Sourcing Decisions,* Washington, D.C., GAO-03-475, 2003.

[10] Defense Science Board Task Force, *Human Resources Strategy,* Washington, D.C.: Department of Defense, 2000.

[11] Office of Management and Budget, *Inherently Governmental Functions,* Policy Letter 92-1, 1992.

[12] Office of Management and Budget, 1992.

[13] Office of Management and Budget, 1992.

[14] Indeed, the 2005 congressionally mandated Acquisition Advisory Panel (AAP) found that "[t]he term 'Inherently Governmental' is inconsistently applied across government agencies" (Acquisition Advisory Panel, 2007).

[15] Cynthia Quarterman, "Creating a Government That Works Better and Costs Less: How Far Have We Come?" *Mineral Management Service Today,* Vol. 6, No. 2, 1996.

to redesign, to reinvent, to reinvigorate the entire national government."[16] One of the hundreds of recommendations issued in the report by Vice President Al Gore was, "Outsource non-core Department of Defense functions." [17] Five years later, the Brookings Institution estimated that, under Clinton's plan, DoD reduced its workforce by 25 percent.[18] Overall, the plan called for downsizing the federal workforce by 252,000 workers.[19]

Congressional action in the 1990s also facilitated outsourcing efforts.[20] In 1998, Congress passed the Federal Activities Inventory Reform Act (FAIR Act),[21] which required agencies to identify inherently governmental and not inherently governmental functions. It also called on agencies to manage competitions to determine whether private-sector or governmental performance was most appropriate.[22] Senator Thomas (R-Wyo.), one of the bill's sponsors, explained that the legislation "would establish a statutory basis for determining whether a good or service from the government could be provided more cost-effectively by the government or the private sector." Furthermore, Senator Thomas explained, "[i]t would establish a preference for the private sector," except in the case of IG functions.[23]

The George W. Bush Administration, 2001–2009

The George W. Bush administration continued existing trends and made significant changes in the use of contractors in federal workplaces. One very prominent development during this period was the widespread use of security contractors in Iraq and Afghanistan, which attracted significant attention from the media and government watchdogs.[24] The Bush administration also made significant changes to the use of contractors at home. In 2002, it released the President's Management Agenda (PMA), a series of initiatives aimed at improving the management and efficiency of government operations, in part through greater reliance on the private sector. One initiative called for increased use of competitive sourcing with the private sector.[25] According to Valerie Bailey Grasso, the PMA asserted that "nearly half of all federal employees perform tasks that are readily available in the commercial marketplace." To address this perceived

---

[16] Al Gore, *From Red Tape to Results: Creating a Government That Works Better & Costs Less*, Washington, D.C.: Office of the Vice President, 1993.

[17] Gore, 1993.

[18] Donald F. Kettl, *Reinventing Government: A Fifth-Year Report Card*, Washington, D.C.: The Brookings Institution, 1998.

[19] Kettl, 1998.

[20] Other acquisition reforms of this period designed to facilitate access to commercial products include the Federal Acquisition Streamlining Act of 1994 and the Federal Acquisition Reform Act of 1996 (Acquisition Advisory Panel, 2007).

[21] Public Law 105-270, Federal Activities Inventory Reform Act of 1998, October 19, 1998.

[22] Grasso, 2005.

[23] *Hearing Before the Subcommittee on Government Management, Information, and Technology of the Committee on Government Reform and Oversight, House of Representatives, One Hundred Fifth Congress, First Session, on H.R. 719*, Washington, D.C.: U.S. Government Printing Office, 1998.

[24] See for example, Congressional Budget Office, *Contractors' Support of U.S. Operations in Iraq*, Washington, D.C., 2008; Moshe Schwartz, *Department of Defense Contractors in Iraq and Afghanistan: Background and Analysis*, Washington, D.C.: Congressional Research Service, 2010.

[25] Grasso, 2005.

underutilization of private contractors, the PMA endorsed the use of competitions to identify functions that could better be performed by the private sector.[26]

In a move indicative of changes in this period in regard to competitive sourcing, the Bush administration also significantly revised OMB Circular No. A-76, "Performance of Commercial Activities." First introduced in 1966 (and subject to subsequent revision), Circular No. A-76 outlines a formal and complex process for conducting "managed competitions" between the private and public sectors. The circular assumes that the federal government should not compete to complete tasks that could more efficiently be performed by the private sector, and it also assumes that efficiency can be determined by rigorous cost competition.[27] In 2002, the Bush administration recommended revisions to Circular No. A-76 intended to establish a presumption "that all activities are commercial in nature unless an activity is justified as inherently governmental."[28] This change was codified in 2003 in a revision that narrowed the definition of *inherently governmental*. Whereas the previous language had defined *inherently governmental* as an activity that requires "the exercise of discretion in applying government authority and/or in making decisions for the government," the 2003 version added the term "substantial" before "discretion."[29] In 2008, Bernard D. Rostker found that this change "reduced the impact of the provision by allowing outsourcing of many activities that were previously considered 'inherently governmental.'"[30]

Rethinking Outsourcing

The scope and consequences of outsourcing activities led to evaluations of and some broad rethinking about the existing approach. In 2008, the Government Accountability Office (GAO) urged the government to adopt a more strategic approach to hiring contractors.[31] It warned that, "unless the federal government pays the needed attention to the types of functions and activities performed by contractors, agencies run the risk of losing accountability and control over mission-related decisions."[32] In a separate report, GAO highlighted the particularly acute challenges faced by the Department of Homeland Security, an agency stood up so quickly that, as of February 2010, it had more contractors than it did federal employees.[33] GAO found that, "until the department emplaces the staff and expertise needed to oversee selected services, it will continue to risk transferring government responsibility to contractors."[34] Department of

---

[26] Grasso, 2005.

[27] Grasso, 2005.

[28] Office of Management and Budget, "Performance of Commercial Activities," *Federal Register*, Vol. 67, No. 223, November 19, 2002.

[29] Bernard D. Rostker, *A Call to Revitalize the Engines of Government*, Santa Monica, Calif.: RAND Corporation, OP-240-OSD, 2008.

[30] Rostker, 2008.

[31] U.S. Government Accountability Office, *Defense Management: DoD Needs to Reexamine Its Extensive Reliance on Contractors and Continue to Improve Management and Oversight*, Washington, D.C., GAO-08-572T, 2008.

[32] U.S. Government Accountability Office, 2008.

[33] Ed O'Keefe, "At Homeland Security, Contractors Outnumber Federal Workers," *Washington Post*, February 25, 2010.

[34] U.S. Government Accountability Office, *Department of Homeland Security: Improved Assessment and Oversight Needed to Manage Risk of Contracting Selected Services*, Washington, D.C., GAO-07-990, 2007.

Homeland Security officials explained the department's pervasive reliance on contractors by citing short-term needs and a lack of staff and expertise.

In a congressionally mandated study, the AAP conducted a broad review of government acquisition laws, policies, and commercial practices.[35] In a chapter entitled "Appropriate Role of Contractors Supporting Government," the AAP report notes that,

> in some cases, contractors are solely or predominantly responsible for the performance of mission-critical functions that were traditionally performed by civil servants . . . . These developments have created issues with respect to the proper roles of, and relationships between, federal employees and contractor employees.[36]

The AAP recommends that OFPP clarify guidance on the division between governmental and nongovernmental functions and that it ensure that IG functions are adequately staffed.

Congress took note of ongoing concerns about existing DoD sourcing decisions. The National Defense Authorization Act for Fiscal Year 2006 required that DoD "prescribe guidelines and procedures to ensure that consideration is given to using federal government employees for work that is currently performed, or would otherwise be performed, under Department of Defense contracts."[37] This requirement was reiterated and somewhat strengthened in the National Defense Authorization Act for Fiscal Year 2008.[38]

In-Sourcing

Although in-sourcing initiatives were promulgated by members of Congress as early as 2006, the incoming Obama administration gave the issue new prominence and priority. The President expressed concern that the scale and scope of contractors had grown to such a degree that the contractors defied effective oversight. He asserted that, between 2001 and 2008, spending on government contracts had more than doubled, reaching more than $500 billion by 2008.[39] It was not just the scale that was troubling; officials also stated that contractor functions were beginning to encompass IG work.[40] Critics also questioned the extent to which anticipated efficiencies from outsourcing did in fact materialize. In a reversal from past assumptions about the substantial cost savings to be gleaned from outsourcing, proponents of in-sourcing now asserted that bringing functions back in house would save the government money. The House

---

[35] This report also offers one of the few extensive discussions on the issue of personal-services contracts. The authors write, "The current prohibition on personal services contracts has forced agencies to create unwieldy procedural safeguards and guidelines to avoid entering into personal service contracts, some of which may cause the administration of the resulting 'non-personal' contracts to be inefficient." The authors also call for an end to the prohibition (Acquisition Advisory Panel, 2007).

[36] Acquisition Advisory Panel, 2007.

[37] Under Secretary of Defense, "Implementation of Section 343 of the 2006 National Defense Authorization Act," memorandum, Washington, D.C., July 27, 2007.

[38] Public Law 110-181, National Defense Authorization Act for Fiscal Year 2008, January 28, 2008.

[39] The White House, "Memorandum for the Heads of Executive Departments and Agencies, Subject: Government Contracting," press release, March 4, 2009.

[40] Matthew Weigelt, "Obama Hits Campaign Trail to Sell In Sourcing," WashingtonTechnology.com, January 12, 2010a.

Appropriations Committee anticipated that the in-sourcing program would save $44,000 per year for each position converted.[41]

Such concerns led the incoming Obama administration to make in-sourcing a top priority. In general, the administration's language characterized in-sourcing initiatives as an attempt to reverse recent trends rather than an articulation of a bias toward government service. In the *Quadrennial Defense Review* released in February 2010, DoD announced plans to return the number of contractors to pre-2001 levels. According to the *Quadrennial Defense Review*, this change "more appropriately aligns public- and private-sector functions, and results in better value for the taxpayer."[42] In March 2010, the President announced his priorities with regard to government contracting. Noting that "government outsourcing for services raises special concerns," President Barack Obama called on OMB to issue new guidance clarifying the line between IG functions and commercial activities.[43] This was an effort to answer the historically tricky question of how to define and identify IG functions. In an April 2009 press briefing, Secretary Gates articulated a recommendation, later reiterated in the *Quadrennial Defense Review*, that the number of support-service contractors be scaled back to pre-2001 levels. "Our goal is to hire as many as 13,000 new civil servants in FY10 to replace contractors and up to 30,000 new civil servants in place of contractors over the next five years," he explained.[44] To support in-sourcing plans, Congress appropriated $5 billion for FY 2010.[45]

In May 2009, the administration followed up its stated in-sourcing goals with guidance on how to implement the effort. Deputy Secretary of Defense William Lynn released a memorandum, "In-Sourcing Contracted Services—Implementation Guidance," that outlined responsible authorities, criteria for in-sourcing decisions, and a flowchart establishing a process for prioritizing and carrying out in-sourcing actions.[46] Eleven members of Congress wrote to Secretary Gates expressing concern over the memorandum and the danger of in-sourcing "too far or too fast." Specifically, they were concerned that flow chart appeared to "show a very strong bias toward in-sourcing."[47] In January 2010, the Director of Cost Assessment and Program Evaluation issued guidance on conducting cost comparisons of civilian, military, and contract support.[48] In the context of the administration's in-sourcing agenda, this memorandum was meant to provide guidance on whether in-sourcing would be cost-effective.[49]

---

[41] Amber Corrin, "DoD Gets Ball Rolling on In-Sourcing," *Defense Systems*, 2010.

[42] U.S. Department of Defense, 2010a.

[43] This guidance was to be issued on March 31, 2010; it is discussed later in this chapter and in Chapter Three.

[44] U.S. Department of Defense, 2009. He repeated this proposal in his May 7, 2009, budget announcement (Luckey, Grasso, and Manuel, 2009).

[45] Corrin, 2010.

[46] Deputy Secretary of Defense, 2009.

[47] Gates, 2010; Robert J., Wittman, Jim Moran, Jeff Miller, Todd Tiahrt, Joe Wilson, J. Randy Forbes, Michael R. Turner, Paul C. Broun, Doug Lamborn, Duncan Hunter, and Bill Posey, letter to Robert M. Gates, Washington, D.C., July 31, 2009.

[48] Office of the Secretary of Defense, "Estimating and Comparing the Full Costs of Civilian and Military Manpower and Contract Support," Directive-Type Memorandum 09-007, January 29, 2010.

[49] Some critics view this guidance as insufficient. Alan Chvotkin, executive vice president and counsel of the Professional Services Council, called it "a rudimentary cost comparison methodology process" that "does not provide a cogent methodology to enable appropriate and consistent implementation" (U.S. Senate Committee on Homeland Security and Governmental Affairs, *Balancing Act: Efforts to Right-Size the Federal Employee-to-Contractor Mix*, Washington, D.C., 2010).

On March 31, 2010, OFPP issued draft guidance on positions suitable for in-sourcing. The authors of this guidance intended to clarify what constitutes an IG function. "There are too many anecdotes that suggest work that is really inherently governmental . . . is, in fact, being done by contractors," stated Daniel Gordon, head of OFPP.[50] Gordon described the guidance as a response to widespread confusion about what functions could and could not appropriately be in-sourced and as an effort to establish "a clear and comprehensive policy framework."[51] The guidance, which is discussed at length in Chapter Three, issued a single definition of *inherently governmental*, clarified which work is considered CAWIG, and issued guidance for the first time on the treatment of "critical functions." It also required agencies to take steps before and after awarding a contract to ensure that performance of functions remains appropriate.

In May 2010, the Senate Subcommittee on Oversight of Government Management, the Federal Workforce, and the District of Columbia held hearings on the status of in-sourcing efforts. The subcommittee heard testimony from personnel policy representatives, governmental auditors, representatives of federal employee unions, and representatives of private-sector groups. Although the hearing revealed enthusiasm for and optimism about in-sourcing initiatives, critics also highlighted implementation challenges.

Pushback Against In-Sourcing Initiatives

A key assertion made by critics of recent in-sourcing initiatives is that decisions about the workforce balance should be made without quotas or bias (whether toward in-sourcing or outsourcing) and on the basis of strategic assessments. Similar concerns have been expressed by critics of outsourcing initiatives. For example, Greg Carlstrom noted in 2010 that "[f]ederal agencies complained for years during the [George W.] Bush administration about what they called an overly broad, poorly designed outsourcing program; now, it seems, federal contractors have the same complaint about President Obama's in-sourcing program."[52] A prominent critic of President Obama's program is Stan Soloway, president and chief executive of the Professional Services Council, an organization of government services contractors. In a July 2010 op-ed in the *Washington Post*, Soloway called the activities under way in DoD a "well-intended workforce initiative . . . devolving into a quota-driven numbers game."[53] He also questioned the basis for projected cost savings and objected to what he characterized as haphazard conversions made with insufficient regard for the nature of functions performed. These were, of course, the same critiques leveled against outsourcing initiatives in previous decades. Yet, concern about the lack of a strategic approach to in-sourcing led the House Armed Services Committee in May 2010 to adopt an amendment to the Ike Skelton National Defense Authorization Act for Fiscal Year 2011 that would prevent using quotas as the basis for in-sourcing decisions. The difficulty of developing a strategic and rigorous approach to sourcing decisions has long plagued both outsourcing and in-sourcing advocates. A pervasive lack of data on contracts, numbers of contractors, and functions actually performed—combined with disagreement about how to

---

[50] Matthew Weigelt, "OFPP Proposes Tests for Deciding When to Outsource Work," WashingtonTechnology.com, March 31, 2010b.

[51] Weigelt, 2010b.

[52] Greg Carlstrom, "Tables Turned: Contractors Complain In-Sourcing Tactics Unfair," FederalTimes.com, last updated March 21, 2010.

[53] Stan Soloway, "Defense Department's Approach to In-Sourcing Has Unintended Consequences," *Washington Post*, July 19, 2010.

conduct cost comparisons[54]—has led to unresolved controversy. In the absence of a clear basis for making such strategic decisions, critics have charged, DoD risks inconsistent outcomes. Indeed, in January 2010, a company that contracted with the U.S. Air Force to provide audio-visual services sued in federal court to challenge a recent decision to in-source. The company filed a Freedom of Information Act request to access the Air Force's cost analysis and characterized the Air Force's rationale as lacking. Several months later, the Air Force reversed its in-sourcing decision and even extended the contract.[55]

Challenges to Implementing In-Sourcing

Structural Factors

In a 2009, authors writing for the Congressional Research Service identified "structural factors prompting agencies to rely on contractors."[56] They characterized "personnel ceilings" as a hindrance or deterrent to hiring civil servants.[57] Although the FAR prohibits the use of contractors for the purpose of bypassing personnel ceilings (a matter discussed further in Chapter Three), contractors have been used for precisely this purpose. Such contracts allow an organization to augment the available workforce without exceeding the number of allotted personnel slots.

Another common concern regarding effective implementation of in-sourcing has been the ponderous process of hiring civil servants.[58] If one of the appeals of hiring contractors in the first place is the fact that they can be hired quickly and flexibly, hiring civil servants appears to present exactly the opposite situation. In hearings in 2010, Senator Daniel K. Akaka (D-Hawaii) noted, "The long and complicated hiring process across the Federal government may encourage agencies to use contractors rather than hiring permanent staff."[59] One way around this challenge is so-called direct-hire authority, which the Department of Homeland Security requested to support its in-sourcing efforts.[60] "We must ensure that the goals we are asking agencies to achieve with respect to in-sourcing," Senator George Voinovich (R-Ohio)

---

[54] Susan M. Gates and Albert A. Robbert, *Personnel Savings in Competitively Sourced DoD Activities: Are They Real? Will They Last?* Santa Monica, Calif.: RAND Corporation, MR-1117-OSD, 2000.

[55] Matthew Weigelt, "Small Business Fights In-Sourcing . . . and Wins," WashingtonTechnology.com, May 5, 2010c.

[56] Luckey, Grasso, and Manuel, 2009.

[57] "A personnel ceiling establishes the maximum number of positions that may be budgeted in a job category or for all personnel in an organization" (Luckey, Grasso, and Manuel, 2009).

[58] In a move indicative of recent frustration with federal hiring, in 2003, DoD identified a process to replace the existing "cumbersome" General Schedule system with the National Security Personnel System; implementation began in 2006 (Wendy Ginsberg, *Pay-for-Performance: The National Security Personnel System*, Washington, D.C.: Congressional Research Service, 2008). However, the new system has been challenged in courts and has only partially been implemented. In March 2009, the Obama administration announced that it was reviewing the new system and that it would temporarily suspend further implementation.

[59] U.S. Senate Committee on Homeland Security and Governmental Affairs, 2010.

[60] "A Direct-Hire Authority . . . enables an agency to hire, after public notice is given, any qualified applicant without regard to 5 U.S.C. 3309-3318, 5 CFR [Code of Federal Regulations] part 211, or 5 CFR part 337, subpart A. It expedites hiring by eliminating competitive rating and ranking, veterans' preference, and 'rule of three' procedures" (U.S. Office of Personnel Management, "Direct-Hire Authority (DHA) Fact Sheet," web page, undated).

noted, "can be achieved using current hiring tools. If not, the Administration or Congress must supply agencies with sufficient flexibilities to get the job done."[61]

Review of Progress to Date

There has been little formal assessment of the progress of in-sourcing in DoD, but reports on the experiences of civilian agencies may offer relevant insights. In response to a congressional mandate to review the status of in-sourcing efforts, GAO reported in October 2009 on the progress of in-sourcing in nine civilian agencies.[62] It noted that, although the civilian agencies had been required to develop and implement in-sourcing guidelines and procedures by July 2009, "[n]one of the nine civilian agencies . . . visited [by October 2009 had] met the statutory date."[63] Officials supplied several reasons for the delay, including unclear guidance and the complexity of the task, which involves broad coordination across the organization and a significant commitment of time. Specifically, officials expressed uncertainty about the meaning of such terms as *inherently governmental, mission-critical, core competency*, and *consideration* (vs. *special consideration*).[64] They also noted a lack of clarity about when a cost analysis is required and how to appropriately conduct one and expressed how difficult it is to gather or analyze certain kinds of contract data that should shape in-sourcing decisions. They also voiced concern about gaps in the workforce that result, once an in-sourcing decision has been made, from the significant time and resources expended to hire a civil servant.[65]

Less than a year later, GAO offered an in-sourcing status update. Its May 2010 report emphasized that effective implementation would require fundamental changes to the way the government manages its resources and assesses progress toward goals.[66] The author, John K. Needham, noted that effective implementation would require that agencies have a sound strategic human-capital plan. GAO identified strategic human-capital management as a "high-risk area" in 2001, but, in 2010, it reported that much progress still needed to be made. To bolster strategic planning, the GAO report recommended inventorying service contracts; developing an analytical approach to assessing whether there is a business case for in-sourcing; and developing more-flexible tools for recruiting, hiring, and managing the workforce.[67]

---

[61] U.S. Senate Committee on Homeland Security and Governmental Affairs, 2010.

[62] The audited agencies were the Department of Energy, the General Services Administration, the Department of Health and Human Services, the Department of Homeland Security, the Department of Justice, the National Aeronautics and Space Administration, the Department of State, the Department of Transportation, and the Department of Veterans Affairs (U.S. Government Accountability Office, *Civilian Agencies' Development and Implementation of In-Sourcing Guidelines*, Washington, D.C., GAO-10-58R, 2009).

[63] U.S. Government Accountability Office, 2009.

[64] U.S. Government Accountability Office, 2009.

[65] U.S. Government Accountability Office, 2009.

[66] John K. Needham, *Sourcing Policy: Initial Agency Efforts to Balance the Government to Contractor Mix in the Multisector Workforce: Testimony Before the Subcommittee on Oversight of Government Management, the Federal Workforce, and the District of Columbia, Committee on Homeland Security and Governmental Affairs, U.S. Senate*, Washington, D.C.: Government Accountability Office, GAO-10-744T, May 10, 2010.

[67] Needham, 2010.

Postscript: In-Sourcing's Uncertain Future

Despite the extent to which to the goal of reducing DoD's dependence on contractors had been enthusiastically embraced at the highest levels, the future of in-sourcing efforts is currently uncertain. In August 2010, Secretary Gates announced a series of "Efficiencies Initiatives" intended to "reform the way the Pentagon does business."[68] Some bear directly on the implementation of in-sourcing efforts. The first calls for a reduction in funding of support contractors by 10 percent in each of the next three fiscal years. The second calls for a freeze at FY 2010 levels in the number of OSD, defense agency, and combatant commander billets (not in the services, however; they are exempt from the billet freeze) for each of the next three fiscal years. To clarify, Secretary Gates noted, "With regard to in-sourcing, no more full-time OSD positions will be created after fiscal [year] 2010 to replace contractors except for critical needs."[69] In-sourcing will be allowed to continue in the military services.[70]

These August 2010 pronouncements have led some observers to view the administration's brief commitment to in-sourcing all but dead: "Pentagon Abandons In-Sourcing Effort," announced a headline in *Government Executive*.[71] "As we were reducing contractors, we weren't seeing the savings we had hoped from in-sourcing," stated the Secretary in a line reported by *The Huntsville Times*.[72] Organizations representing the interests of private firms applauded the announcement, but unions of government employees decried the move.[73] Amid competing priorities for time and in an era of scarce resources, it seemed that the significant effort required to effect change in the composition of the DoD workforce had fallen by the wayside. Yet, by September 2010, there were indications that reports of in-sourcing's demise had been premature. For example, DoD explained that, in FY 2011, the military departments intended to proceed with in-sourcing as previously planned.[74]

The ongoing battle over the future of in-sourcing was particularly conspicuous in a September 2010 House Budget Committee hearing on the Pentagon's efficiency efforts.[75] Representatives of private contractors and civil-service unions clashed at the hearing, reflecting the reality that powerful stakeholders continue to struggle to shape sourcing policy.[76] Soloway reiterated his concerns about in-sourcing "quotas" and the insufficient analytical basis used to make conversions. However, Jacqueline Simon of the American Federation of Government Employees retorted that

---

[68] Office of the Assistant Secretary of Defense (Public Affairs), 2010.

[69] Office of the Assistant Secretary of Defense (Public Affairs), 2010.

[70] Robert Brodsky, "Union Blasts Scaled Down Defense In-Sourcing Plan," September 30, 2010c.

[71] Robert Brodsky, "Pentagon Abandons In-Sourcing Effort," GovernmentExecutive.com, August 10, 2010a.

[72] Kenneth Kesner, "Defense Secretary Says In-Sourcing Hasn't Cut Costs as Hoped; Future of Initiative Uncertain," *The Huntsville Times*, August 29, 2010.

[73] Brodsky, 2010a.

[74] Robert Brodsky, "Defense In-Sourcing to Continue at Military Services," GovernmentExecutive.com, September 7, 2010b.

[75] Jacqueline Simon, "Statement Before the House Budget Committee on the Department of Defense Efficiency Initiative," September 30, 2010.

[76] Brodsky, 2010c.

[c]ontractors, without complaint, took tens of thousands of federal employee jobs during the previous two Administrations, without ever having to compete for our work . . . . [I]f outsourcing had been held to the same standards that contractors would like to apply to in-sourcing, outsourcing would have been suspended no later than close of business on the first day of the Republic.[77]

In-sourcing decisions remain fraught with such issues as good governance, efficiency, public interest, and the role of the private sector in public life. The future of in-sourcing and outsourcing policies will likely be as turbulent as in the past.

Summary

DoD sourcing's recent history has been dominated by the department's belief that increasing contractor performance of government functions could increase DoD efficiencies and mission performance. Clinton- and George W. Bush–era policies expanded opportunities for the private sector in government. But, by the mid-2000s, governmental reviews began calling into question the transparency, accountability, and cost-effectiveness of prior outsourcing decisions, and Congress introduced in-sourcing in the National Defense Authorization Act for Fiscal Year 2006. The incoming Obama administration supported in-sourcing initiatives, giving the issue new urgency and prominence. The administration believed that the scope of contractors had grown to such a degree that contractors defied effective oversight. Officials stated that contractor functions were beginning to encompass IG work.[78] Critics of outsourcing also questioned the extent to which anticipated efficiencies from outsourcing had actually materialized.

Interestingly, in-sourcing and outsourcing have been subject to similar critiques. Detractors have noted the difficulty of developing a strategic and rigorous approach to sourcing decisions; the level of interpretation required to determine what functions are inherently governmental; and a pervasive lack of data on contracts, numbers of contractors, and functions actually performed. Furthermore, disagreement about how to conduct cost comparisons has led to unresolved controversy. Powerful representatives of private contractors on the one hand and civil-service unions on the other continue to struggle to shape sourcing policy in their constituents' favor. However, pushing the pendulum toward in-sourcing has proved significantly more difficult than moving toward outsourcing. The ease and speed of hiring contractors makes outsourcing appealing, and in-sourcing efforts (i.e., working within existent civil-service laws to hire contractors as government employees) are considerably more difficult and therefore less appealing. With no clear resolution of these challenges, the future of sourcing policy is likely to be as turbulent as its past.

---

[77] Simon, 2010.

[78] Weigelt, 2010a.

CHAPTER THREE

Overview of Current Policy and Guidance

Introduction

The history of DoD contracting and the periodic shifts between in-sourcing and outsourcing goals are reflected in law and policy. Statutes, regulations, and executive agency policies have at times condoned and at times prohibited government use of private-sector employees. On the one hand, the government, by law, must ensure that federal employees—and only federal employees—perform core government functions, such as conducting criminal hearings and diplomatic missions. On the other hand, day-to-day activities, such as facilities maintenance and database management, could potentially be procured at lower cost from the private sector. To secure these cost savings, the government is required to inventory and publish a list of non-core functions that are appropriate for procurement through the private sector.[1] The task is to determine which functions are appropriate for contractor performance and which are not.

This chapter presents an overview of existing policy and guidance, and it briefly touches on the ongoing revision to core terms and definitions. Unless otherwise stated, the information presented was current as of June 2010.[2] The information in this chapter formed the basis of the development of our analytical approach to assessing in-sourcing. We describe four major restrictions that govern the use of contractors, and we identify pending and proposed changes to law and policy, where appropriate.

Four Major Rules Restrict the Use of Contractors

Law and policy establish restrictions on the use of contractors for performance of functions that should be the exclusive or near-exclusive responsibility of government employees. These general concepts identify the nature of work that contractors may not perform or that they may perform only under specific circumstances. There are four major restrictions on the use of contractors:

- a prohibition on contractor performance of IG functions
- special rules about the use of contractors to perform CAWIG functions
- a prohibition on the use of personal-services contracts

---

[1] See Public Law 105-270. This chapter discusses the FAIR Act in detail in the next section.

[2] Writing was complete as of June 2010; revision, especially of the references, continued through December 2010.

- DoD-specific exemptions from private-sector performance of specific commercial functions identified in law and policy.

As of this writing, there is also a proposed new restriction on contractor performance of critical functions.

This section identifies the sources of these restrictions, describes their requirements, and provides a brief update on currently proposed policy that may change the existing regulations. These concepts constitute both the basis of DoD's policy governing the use of contractors and the main regulatory material for assessment of a contractor workforce.

No Contractor Performance of Inherently Governmental Functions

Contractors should never perform IG functions. This notion is central to the government's approach to regulating the use of contractor support.[3] There are several definitions of IG functions in law and in policy,[4] and these definitions are circular and only implementable by example.

DoDI 1100.22, *Policy and Procedures for Determining Workforce Mix*, contains DoD's policy for IG functions. It states that IG functions include, among others,

> activities that require either the exercise of substantial discretion when applying Federal Government authority; or value judgments when making decisions for the Federal Government, including judgments relating to monetary transactions and entitlements.[5]

The general idea is that government employees alone should make government decisions. Although "substantial discretion" is a term of art, we interpret the policy as requiring contracting officials to distinguish between decisionmaking on the one hand and actions in support of decisionmaking on the other. If the decision is of consequence to a program, acquisition system, policy, or business model, for example, the decision should be made by a government employee.

DoD policy is based on three other documents that help us understand the ban on contractors performing IG functions: the FAIR Act, OMB Circular No. A-76, and the FAR. These documents provide some clarity about the meaning of IG functions. We review each of them in turn to provide a deeper understanding of the IG function restriction.

---

[3] For an excellent and thorough review of the issue, see Luckey, Grasso, and Manuel, 2009.

[4] For example, 10 U.S.C. 2383 defines IG functions by referring to a General Services Administration regulatory definition contained in the FAR (see 48 C.F.R. § 7.5 and the definition of *inherently governmental function* at 48 C.F.R. § 2.101. The latter notes that the FAR definition is "a policy determination, not a legal determination"). The FAR reference imperfectly reproduces an earlier definition promulgated by OMB in an instruction to all federal agencies (Office of Management and Budget, Circular No. A-76 Revised, May 29, 2003).

The 1998 FAIR Act provides yet another definition of *inherently governmental functions*. There is at least one other statutory definition (which is reproduced in two separate acts of Congress), several statutory references (which declare certain government functions to be inherently governmental), and a host of policy definitions. See, for example, 5 U.S.C. 306 (which declares the drafting of agency strategic plans to be an IG function that must be performed by federal employees); 33 U.S.C. 2321 (which states that the operation and maintenance of U.S. Army Corp of Engineers' hydroelectric power plants is an IG function); and Public Law 110-28, U.S. Troop Readiness, Veterans' Care, Katrina Recovery, and Iraq Accountability Appropriations Act, 2007, May 25, 2007 (which declares that employee actions at the National Energy Technology Laboratory are IG functions).

[5] U.S Department of Defense, 2010b.

The FAIR Act, passed in 1998, was an attempt to reap cost savings by outsourcing commercial activities. It requires the heads of each executive agency, including DoD, to publish a list of "activities performed by Federal Government sources for the agency that, in the judgment of the head of the executive agency, are not inherently governmental functions."

The FAIR Act defines an IG function as "a function that is so intimately related to the public interest as to require performance by Federal Government employees." To supplement the definition, the FAIR Act provides a list of representative functions that Congress considers inherently governmental, including functions that

- bind the United States to take or not to take some action by contract, policy, regulation, authorization, order, or otherwise;
- determine, protect, and advance United States economic, political, territorial, property, or other interests by military or diplomatic action, civil or criminal judicial proceedings, contract management, or otherwise;
- significantly affect the life, liberty, or property of private persons;
- commission, appoint, direct, or control officers or employees of the United States; or
- exert ultimate control over the acquisition, use, or disposition of the property, real or personal, tangible or intangible, of the United States, including the collection, control, or disbursement of appropriated and other Federal funds.[6]

The second document is OMB Circular No. A-76. Circular No. A-76, an executive instruction sent to federal agency heads, reaffirms that the government will rely on commercial sources when private industry can produce the goods and services at the lowest cost.[7] However, it notes that "certain functions are inherently Governmental in nature, being so intimately related to the public interest as to mandate performance only by Federal employees."[8]

Circular No. A-76's definition of IG functions is substantially similar to the FAIR Act's definition, but the circular adds two categories of IG functions: (1) acts of governing and (2) monetary transactions and entitlements. Acts of governing include conducting criminal investigations, managing and directing the armed services, directing federal employees, and selecting program priorities. Monetary transactions and entitlements include tax collection, control of Treasury accounts and money supply, and the administration of public trusts.

The third document is the FAR, subpart 7.5. The FAR does not define IG functions; rather, it provides many examples of IG functions. This extensive list allows manpower planners to reason by analogy: Functions similar to those on the FAR list can logically be consid-

---

[6] Note that the FAIR Act also excludes some functions from the definition of IG functions, including "gathering information for or providing advice, opinions, recommendations, or ideas to Federal Government officials" and ministerial activities, such as building security and conducting mail operations.

[7] Office of Management and Budget, Circular No. A-76, Revised 1999, August 4, 1983. Appendix A of the circular lists examples of commercial activities, ranging from laundry services and library operations to payroll services and printing. Also included are "Special Studies and Analyses," such as cost-benefit analyses, statistical analyses, scientific data studies, regulatory studies, defense studies, legal studies, and management studies.

[8] The text we quote here almost exactly reproduced the definition of *inherently governmental* provided in the Circular: "An inherently Governmental function is a function which is so intimately related to the public interest as to mandate performance by Government employees." Although it is worded differently, the circular's definition of IG functions expresses the same idea as the FAIR Act's definition.

ered IG functions. The FAR's list of IG functions is reproduced in its entirety in Appendix A of this report.

These three documents—the FAIR Act, OMB Circular No. A-76, and the FAR—form the basis of the DoD policy contained in DoDI 1100.22. Contracting officials and manpower planners at DoD have substantial discretion, subject to OMB review, to determine whether a function is inherently governmental and therefore cannot be performed by contractors.

We note that the definition of what constitutes an IG function has come under review and may soon be revised. In the Duncan Hunter National Defense Authorization Act for Fiscal Year 2009 (2009 NDAA), section 321,[9] Congress directed OMB to analyze uses of the term *inherently governmental functions* and then develop a single, government-wide definition.

OFPP released an unofficial draft policy letter that responds to the 2009 NDAA with proposed revisions to the regulation of IG functions.[10] The new policy would define an IG function as "a function so intimately related to the public interest as to require performance by Federal Government employees." The draft policy instructs agencies to evaluate functions on a case-by-case basis according to two tests:

- Nature-of-the-function test: Sovereign power is by definition inherently governmental. Any functions exercising sovereign power—such as diplomatic relations, police authority, and criminal sentencing—must be performed by a government employee.
- Exercise-of-discretion test: Decisions that bind the government to a course of action are inherently governmental. A government employee must perform any function that involves the substantial exercise of discretion that may bind the government to one of several courses of action.

Accordingly, if the agency determines that a function exercises sovereign power or binds the government to a course of action, then that function is considered inherently governmental. The draft policy also details agency and contracting official responsibilities for documenting the position review and identifies processes to be used if and when it is discovered that a contractor is performing IG functions.

The new draft policy is not yet official, and its future is uncertain. But the likely effect will be to reinforce the existing rule: Federal government employees, and only federal government employees, may wield the authority of the federal government.

CAWIG Functions Warrant Special Consideration

Even if a function is not inherently governmental, there may be sound policy reasons for preventing its performance by contractors. For example, although they are not inherently governmental, some functions, such as acquisition support services, may become so entangled with government acquisition decisionmaking as to justify their performance by government employees. To account for this possibility, 10 U.S.C. 2383 governs contractor performance of acquisition functions that are CAWIG.[11] Whereas IG functions should never be outsourced,

---

[9] Public Law 110-417, Duncan Hunter National Defense Authorization Act for Fiscal Year 2009, October 14, 2008 (as amended by Public Law 111-84, National Defense Authorization Act for Fiscal Year 2010, October 28, 2009).

[10] A copy of the notice is available at Office of Management and Budget, Office of Federal Procurement Policy, undated.

[11] Section 2383 was added per Public Law 108-375, Ronald W. Reagan National Defense Authorization Act for Fiscal Year 2005, October 28, 2004.

outsourcing CAWIG functions is allowed in special circumstances. To approve a contract for a CAWIG function, the contracting official must ensure that each of the following is true:

- DoD military or civilian personnel cannot reasonably be made to perform the function.
- Appropriate DoD military or civilian personnel are available both to supervise contractor performance and to perform all related IG functions.
- There are no contractor conflicts of interest.[12]

Here, again, the law is supported by the FAR. The FAR, subpart 7.503(d), does not define CAWIG functions, but it notes that

> certain services and actions that are not considered to be inherently governmental functions may approach being in that category because of the *nature of the function, the manner in which the contractor performs the contract, or the manner in which the Government administers contractor performance.* (emphasis added)

In addition to this justification, the FAR provides a list of representative examples of CAWIG functions. (This list is reproduced in its entirety in Appendix B of this report.) Because the FAR list is nonexhaustive, implementation of the CAWIG rule is likely to depend on reasoning by analogy.

A separate section of U.S. Code, 10 U.S.C. 2463, requires the Under Secretary of Defense for Personnel and Readiness (USD/P&R) to prescribe regulations governing the use of DoD *civilian* employees who perform DoD functions.[13] For example, DoD policy must provide "special consideration" for using civilian DoD employees when there is a contractor performing a CAWIG function (as defined in 10 U.S.C. 2383, which is the FAR definition).[14] In response to 10 U.S.C. 2463, USD/P&R issued a memorandum in 2008 that contains guidance for implementing the special consideration requirement.[15] Rather than stating a preference for civilian performance of CAWIG functions because of the nature of the function, the USD/P&R guidance calls for an economic analysis "to determine whether DoD civilians or private sector contractors are the low cost provider and should perform the work."[16]

The USD/P&R guidance raises a new issue that does not appear to be well defined or understood. The law (10 U.S.C. 2383) seems to imply a preference for civilian employees over contractors when CAWIG functions are involved. Contractors can be used only if there are no

---

[12] In a separate statutory section, 10 U.S.C. 2330a, the law requires the Secretary of Defense to inventory DoD's use of contractors and ensure that (1) the use of contractors for personal services conforms to statutory restrictions, (2) contractors are not performing IG functions (as defined in 10 U.S.C. 2338(b)(2)), and (3) to the maximum extent possible, contractors are not performing CAWIG functions (as defined in 10 U.S.C. 2383).

[13] Section 2463 was added to U.S. Code per the National Defense Authorization Act for Fiscal Year 2008.

[14] Public Law 111-117, Consolidated Appropriations Act, 2010, December 16, 2009, Section 743, requires the heads of executive agencies to review the agency's inventory of contractor functions and ensure that "the agency is giving special management attention to functions that are closely associated with inherently governmental functions" (31 U.S.C. 501, note).

[15] Deputy Secretary of Defense, "Implementation of Section 324 of the National Defense Authorization Act for Fiscal Year 2008 (FY 2008 NDAA)—Guidelines and Procedures on In-Sourcing New and Contracted Out Functions," memorandum, April 4, 2008.

[16] The emphasis on the low-cost provider is consistent with 10 U.S.C. 129a's requirement to use the least-costly form of support.

government employees available to perform or supervise the work. But, as a matter of USD/P&R policy, CAWIG functions should be assigned to the low-cost provider for purely economic reasons. Contractors may be the low-cost provider, regardless of government-employee availability. Should a low-cost contractor be chosen, even if there is government-employee availability? The law is unclear on this point. A further complicating issue is that 10 U.S.C. 2330a requires the Secretary of Defense to ensure, "to the maximum extent possible," that the DoD inventory of contractor functions does not include any CAWIG functions. The tension between this "maximum extent possible" language, the civilian preference required by 10 U.S.C. 2383, and the "low cost provider" language of the USD/P&R guidance appears to have been overlooked, and it still exists in current policy.[17]

No Contracts for Personal Services

Federal rules restrict the use of contractors for "personal services." The general prohibition is contained in the FAR, subpart 37.104.[18] This subpart states that a "personal services contract is characterized by the employer-employee relationship it creates between the Government and the contractor's personnel."[19] To determine whether a contract calls for performance of personal services, the subpart requires "each contract arrangement [to] be judged in the light of its own facts and circumstances, the key question always being: *Will the Government exercise relatively continuous supervision and control over the contractor personnel performing the contract?*" (emphasis added).

DoDI 1100.22 establishes DoD's policy on use of personal-services contracts. Although the general concept it presents is simple, its implementation is nuanced. The department instructs contracting officials evaluating an existing or proposed contract to ask the following questions:

- Is government supervision required in order "to adequately protect the government's interest; retain control of the function involved; or retain full personal responsibility for the function supported in a duly authorized Federal officer or employee"?
- Does the contractor's service directly apply to integral DoD efforts that further assigned functions or missions?
- Are comparable services for comparable needs performed by military or DoD civilian employees?
- Is the service performed on-site, with government-furnished equipment?
- Can the service reasonably be expected to last more than one year?

Our interpretation is that the policy is asking evaluators to judge whether a reasonable person would assume that the contractor performing the work is actually a government employee. We believe that, if there is relatively continuous supervision of that employee and of

---

[17] DoDI 1100.22 continues the confusion: "DoD Components shall use civilian personnel to perform [non-IG and nonexempt] functions unless DoD civilians are not the low-cost provider or there is a legal, regulatory, or procedural impediment to using DoD civilian personnel."

[18] According to the FAR, subpart 37.104(b), "Agencies shall not award personal services contracts unless specifically authorized by statute (e.g., 5 U.S.C. 3109) to do so."

[19] It further explains that "the Government is normally required to obtain its employees by direct hire under competitive appointment or other procedures required by the civil service laws. Obtaining personal services by contract, rather than by direct hire, circumvents those laws unless Congress has specifically authorized acquisition of the services by contract."

his or her workload and work product, and if he or she is performing integral tasks at a government site and with government equipment, an employer-employee relationship can be reasonably assumed. The policy requires an ad hoc analysis that takes all available information about the relationship between the government and the contractor into account.

Temporary Contracts for Advisory and Assistance Services Are Allowed

There is an exception to the ban on personal-services contracts. 10 U.S.C. 129b specifically authorizes DoD to procure personal services from experts and consultants when such advice cannot be obtained from internal sources.[20] The FAR and Defense Federal Acquisition Regulation (DFAR) interpret this law to allow DoD access to specialized knowledge in certain limited cases.

The FAR, subpart 37.2, allows all government agencies to acquire specialized advisory and assistance services. According to the FAR, contracts for advisory and assistance services (CAAS) may be used to

- Obtain outside points of view to avoid too limited judgment on critical issues;
- Obtain advice regarding developments in industry, university, or foundation research;
- Obtain the opinions, special knowledge, or skills of noted experts;
- Enhance the understanding of, and develop alternative solutions to, complex issues;
- Support and improve the operation of organizations; or
- Ensure the more efficient or effective operation of managerial or hardware systems.

The DFAR supplement, subpart 237.104, establishes guidance for expert services and advice.[21] Contracts for expert and consultant services require departmental determination and findings that establish that

- the nature of the contract is temporary or intermittent
- the services are advantageous to national defense
- DoD personnel cannot provide the required skill
- an excepted appointment cannot be obtained
- a non–personal-services contract is impractical
- all statutory restrictions have been met.

The DFAR makes it clear that CAAS are a limited exception to the personal-services ban. Outside technical knowledge can be acquired, but only on a limited, temporary basis.

There is some evidence that CAAS are not well understood and may be used in unintended ways. A 1997 General Accounting Office report revealed that, at the time, DoD likely was underreporting its expenditures on advisory and assistance services. The authors of the report speculated that "the underreporting may be due to difficulties in accurately identi-

---

[20] 10 U.S.C. 129b also permits the Secretary of Defense to use personal-service contracts if the services

are to be provided by individuals outside the United States, regardless of their nationality, and are determined by the Secretary to be necessary and appropriate for supporting the activities and programs of the Department of Defense outside the United States; directly support the mission of a defense intelligence component or counter-intelligence organization of the Department of Defense; or directly support the mission of the special operations command of the Department of Defense.

[21] The DFAR, subpart 237.104, cites the authority for such contracts as Public Law 101-165, Department of Defense Appropriations Act, 1990, November 21, 1989, Section 9002.

fying advisory and assistance task[s]. Several officials observed that the definition of advisory and assistance services was ambiguous, particularly for services related to research and development."[22]

Given both DoD's need for technical assistance and potential confusion about the role of CAAS and the personal-services ban, it is possible that DoD requires clearer guidance on the proper role of CAAS and clearer procedures for acquiring expert services.

Other Functions Are Exempt from Contractor Performance

In general, commercial activities—functions that are neither deemed inherently governmental nor covered by the CAWIG rules—are eligible for contractor performance if the private sector provides the service at the lowest cost.[23] There are, however, limitations on this general rule. Exemptions from contractor performance arise from law, executive order, treaty, and international agreement. We have not researched the raft of exemptions that can apply. Contracting officials are required to certify that any given contract for commercial activities does not run afoul of the exemptions. These assessments are performed on a contract-by-contract basis.

DoD has its own exemptions beyond those that may be contained in law, executive order, treaty, and international agreement. DoDI 1100.22 states that commercial activities may be exempt from contractor performance in the case of

- DoD in-theater readiness needs
- esprit de corps
- overseas, sea-to-shore, and civilian/military rotations
- civilian and military career development
- operational risk
- continuity of operations
- dual-tasked manpower during wartime assignments
- other DoD management decisions.

Per the Deputy Secretary of Defense's memorandum of May 28, 2009, the requiring official must work with the manpower official to provide documentation that the requested functions are not exempt as a matter of law or policy.[24] The memorandum requires expeditious in-sourcing of contracted functions that are subsequently determined to be exempt from private-sector performance.

Potential New Policy and Guidance About Critical Functions

The four elements described in this section—IG functions, CAWIG functions, personal services, and exemptions from private-sector performance—constitute the current core restric-

---

[22] U.S. General Accounting Office, *Defense Advisory and Assistance Service Contracts*, Washington, D.C., B-276026, June 13, 1997.

[23] 10 U.S.C. 129a requires DoD to use the least costly form of support, subject to the additional regulations outlined in this report. In addition to the restrictions described in this section, contracts that have been poorly performed, either for budgetary reasons or in terms of quality, will also trigger special consideration for in-house performance in the future (Public Law 110-417).

[24] Deputy Secretary of Defense, 2009.

tions on the use of contractors. But because many of these restrictions are vague or difficult to apply, the policy debate continues to refine and reconsider appropriate regulation.

Congress has instructed OMB to evaluate the government's use of contractors for functions that are critical to agency missions and operations. Overreliance on the private sector for key but non-IG functions might lead to a lack of in-house capacity. Congress has also instructed OMB to consider a new restriction on government use of contractors for "critical functions." Critical functions are those that, although they are not inherently governmental, are so essential to the agency's mission that the agency must ensure a sufficient government workforce rather than rely too heavily on contractor support. Draft OMB guidance written in response to this tasking sheds light on the new legislative interest in critical functions. The draft letter defines a critical function as "a function that is necessary to the agency being able to effectively perform and maintain control of its mission and operations. A function that would not expose the agency to risk of mission failure if performed entirely by contractors is not a critical function."[25] Government employees must perform critical functions insofar as a sufficient internal workforce is required to maintain agency control over its mission and operations. Agencies with sufficient internal capacity may then use either federal employees or contractors for additional work.[26] If both federal employees and contractors are eligible to perform the work, agencies should perform a cost analysis to determine which source would provide the necessary service at the lowest cost.

Determining what is and is not a critical function is likely as difficult as determining what is and is not an IG or CAWIG function. The core consideration when determining whether contractor performance is appropriate is whether there is sufficient internal capacity to ensure that the agency can maintain control over its missions and operations. This means that there must be an adequate supply of civilian or military employees to meet critical DoD needs and to supervise the supporting contractor base. Just as is required in the case of IG and CAWIG function determinations, the OMB draft policy requires an assessment of critical functions on a case-by-case basis according to the following (nonexclusive) list of factors:

- the agency's mission
- the complexity of the function
- current in-house strength, in terms of capability and capacity
- in-house organic technical expertise
- the effect of contractor default on mission performance
- the enforceability of criminal sanctions for crimes performed by contractors vis-à-vis the laws applicable to federal employees.

The focus on the importance of agency functions in the proposed regulation harkens back to the tension between, on the one hand, the CAWIG-related low-cost provider language of 10 U.S.C. 129a and the USD/P&R implementation guidance and, on the other, the FAR justification for CAWIG restrictions, which are based on the nature of the function or on the manner in which it is performed. The USD/P&R guidance stressed using economic reasons for determining whether or not a function that is considered CAWIG should be in-sourced;

---

[25] A copy of the notice is available at Office of Management and Budget, Office of Federal Procurement Policy, undated.

[26] Initial, immediate guidance was published in Peter R. Orszag, "Managing the Multi-Sector Workforce," memorandum, July 29, 2009.

if it is economical, then it should be in-sourced. The new proposed regulation would swing back closer to the FAR justification, which states that a sufficient DoD workforce must be constantly available before contractor support may be used.

It is important to note that, although this draft letter was written in response to congressional tasking, the new critical-function regulation it advances is merely proposed policy. The public comment period for the draft regulation ran from March 31, 2010, until June 1, 2010, and OMB has yet to enter the next stage in the policy development process. Given the clear congressional interest in critical functions, it is likely that the topic will continue to be the subject of rulemaking.

Summary

Although there are difficulties associated with applying government policy that restricts the use of contractors, that policy does exist. At no time should a contractor be performing work that is inherently governmental. IG functions are so intimately related to the public interest that the government has a duty to perform the work itself. Likewise, the government must not establish an employer-employee relationship with its contractor workforce. If the government wishes to maintain relatively continuous supervision and control over the tasks it needs performed, then the government must rely on the civilian and military workforce.

This does not mean, however, that the government must be overly limiting in its acquisition of outside services. Contracts for expert advice and opinions are a legitimate way to improve policy and decisionmaking. These contracts must be temporary and specialized, though, and they must be used only when similar services are not available from government sources.

Likewise, the government is encouraged to procure commercial goods and services from private-sector sources when they can do so at the lowest cost. Even in such cases, however, there are some exemptions that prohibit contractor performance in the name of overarching policy considerations.

The core restrictions on the government's use of contractor support are being revised to clarify, standardize, and perhaps improve the regulatory environment. But, despite the proposed changes to regulations, the general restrictions likely will remain in place.

CHAPTER FOUR

Methodology

In Chapter Three, we outlined the statutory and policy environment governing contractor performance of government work. Contracting officials need to comply with the law and policy both before a function is tendered for public-private competition and throughout the life of the contract, making sure that a contractor's functions do not creep beyond the terms of the original contract to encompass, for example, IG functions. Given the Obama administration's preference for in-sourcing many previously contracted activities, and given DoD's instructions to the services to "scale back the role of contractors in support services,"[1] the logical starting place for implementing the policies of the President and the department is existing law and policy guidance. It is our interpretation of this guidance that any contracted functions that have strayed from the federal guidelines should be in-sourced.

We developed an analytical approach to help government departments implement in-sourcing plans that comply with federal restrictions on the use of contractors. Our goal was to develop a framework for assessing contractor functions against the federal restrictions to identify instances when contractors might be performing functions that are reserved for government employees. Developing the framework required interpretation of laws, policy, in-sourcing criteria, and contractor functions. Implementing it requires an assessment of the particular mission, roles, and tasks of the organization being reviewed. This chapter presents the approach we developed and describes our interpretations of law and policy. Because law and policy pertaining to IG functions and personal-services contracts are so important, we focused our analysis on these areas. We also developed a questionnaire that allows users to assess these and other criteria. The questions we formulated are based on our interpretation of the FAR, the FAIR Act, and OMB Circular No. A-76. This approach is not a zero-based review, and it does not include cost assessments. The latter are not required for reviews of IG and personal-service functions, which are the focus of this effort.

Analytical Approach

Our analytical approach is based on the 2009 Deputy Secretary of Defense memorandum entitled "In-Sourcing Contracted Services—Implementation Guidance." This guidance was intended to outline a process for determining whether a function should be in-sourced. With some minor modifications, this process is shown in Figure 4.1.[2]

---

[1]   Deputy Secretary of Defense, 2009.

[2]   Note that the memorandum requires a cost analysis only in the case of problems with contract administration or for special-consideration assessments.

Figure 4.1
The In-Sourcing Assessment Process

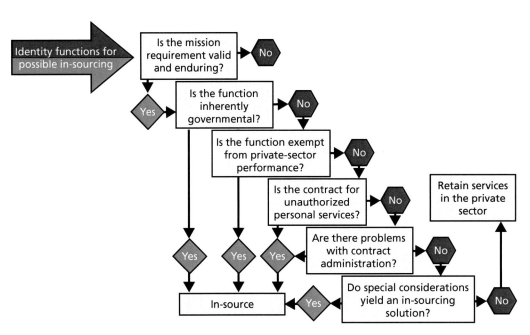

SOURCE: Based on Deputy Secretary of Defense, 2009.
RAND *TR944-4.1*

The process is essentially a step-by-step schematic for determining whether, according to federal rules, a contracted function should be in-sourced. This process was the basis of our analytical approach. Our task was to take the general schematic and derive an analytical method for reviewing specific contractor functions.

To apply the process to a population of contractors, we first adopted from existing guidance and instructions definitions of the following: *IG functions, personal services, exemptions from private-sector performance,* and *contractual problems.*[3] Although this step does remove subjectivity, it enables assessments based on constant definitions. As discussed in Chapter Three, these terms are sometimes defined in several different ways, in multiple pieces of legislation, and in vague terms that are constantly being revised. We chose definitions based on our interpretation of relevant law and policy at the time our study was conducted. Our definitions of these main evaluation criteria are provided in Table 4.1.

---

[3]   See Chapter Three for our review of existing law and policy. Note that this list contains a node in the process that we have not previously discussed: problems with contract administration. The FAR, subpart 37.102(h), allows in-sourcing of a contracted service if there is insufficient in-house staff for the proper management and oversight of the contract. This is not, per se, a restriction on the use of contractors. It is, however, an administrative guide that allows (but does not mandate) in-sourcing of contracted functions that are insufficiently managed or overseen. This is a judgment call to be made by the contracting officer. We did not fully assess the topic of evaluation of contract management, but we wish to acknowledge that this issue is treated in in-sourcing guidance.

Table 4.1
Definitions for Evaluation Criteria

Criteria	Definition (Source)	Is In-Sourcing Automatic, or Does It Warrant Special Consideration?
Inherently governmental	"An activity so intimately related to the public interest as to mandate performance by Federal employees." (FAIR Act)	Automatic
Personal-services contract	"A contract characterized by the employer-employee relationship it creates between the Government and the contractor's personnel—contracts for which the Government exercises relatively continuous supervision and control over the contractor." (FAR)	Automatic
Exemptions from private-sector performance	"Certain specific functions are exempt from private sector performance to provide for the readiness and workforce management needs of the DoD." (DoDI 1100.22)	Automatic
Problems with contract administration	"Problems arising when there are inadequate numbers of trained and experienced officials available within the agency to manage and oversee contract administration." (FAR)	Automatic
Closely associated with inherently governmental	"Functions which approach the status of inherently governmental work because of the nature of these functions and the risk that their performance, if not appropriately managed, may materially limit Federal officials' performance of inherently governmental functions." (Office of Management and Budget, Office of Federal Procurement Policy, undated)	Special consideration

Sources of Information for Making In-Sourcing Assessments

To assess the applicability of existing in-sourcing policy to current functions, we developed a questionnaire (which is reproduced in Appendix C of this report). This questionnaire asks respondents (i.e., DoD officials) about the functions performed by their contractors. The questionnaire asks specific questions to gauge whether these functions are inherently governmental, constituted a personal service, or have any other characteristics that would warrant in-sourcing. Our questionnaire uses language taken directly from existing law and policy. For example, because the FAR, subpart 7.5, identifies "the direction and control of Federal employees" as an IG function, the questionnaire asks respondents whether a contractor's functions include "commissioning, appointing, directing or controlling officers or employees of the U.S." Our questionnaire is designed to help the in-sourcing analyst arrive at "yes" or "no" answers, in accordance with DoD in-sourcing guidance, to the questions presented in the decision tree. The questionnaire is also designed to collect narrative descriptions of the functions performed by each contractor.

In some cases, it is difficult to determine how to interpret key terms and how the terms apply to a specific function. Implementing existing guidance requires interpretation and judgment on the part of the in-sourcing official, as well as intimate knowledge of the organization and the actual work being performed. It is quite possible that two analysts, using the same guidance and possessing the same knowledge of the organization, could arrive at different recommendations.

In an attempt to develop a repeatable analytical process, we adopted a specific rubric for determining the applicability of the in-sourcing criteria. We describe the rubric adopted for each of the criteria in this section.

Inherently Governmental

We determined that a function is inherently governmental if any of the following actions are involved:

- binding the United State to take or not to take action by contract, policy, regulation, authorization, order, or otherwise
- determining, protecting, and advancing U.S. economic, political, territorial, property, or other interests by military or diplomatic action, whether contract management action or otherwise
- making decisions that significantly affect the life, liberty, or property interests of private persons
- commissioning, appointing, directing, or controlling officers or employees of the United States
- exerting ultimate control over the acquisition, use, or disposition of the real or personal, tangible or intangible, property of the United States, including the collection, control, or disbursement of appropriated and other federal funds
- determining agency policy, such as determining the content and application of, for example, regulations
- determining federal program priorities or budget requests
- directing and controlling federal employees
- determining what supplies or services are to be acquired by the government
- determining budget policy, guidance, and strategy
- drafting congressional testimony, responses to congressional correspondence, or agency responses to audit reports from the Inspector General, GAO, or another federal audit entity.

As demonstrated in Chapter Three, guidance on the assessment of the IG criterion relies heavily on examples. Our rubric adopted a list of qualifying examples from OMB Circular No. A-76 and from OFPP's proposed policy letter of March 31, 2010. Our questionnaire could also be tailored to a specific organization. (For example, an organization with a budget-development mission may not need to consider questions about the conduct of criminal investigations.) The first four criteria in the list come directly from OMB Circular No. A-76's definition of what constitutes IG functions. The remaining seven examples come from the OFPP policy letter's list of examples of IG functions.

Responses to the questionnaire can be used to assess whether a function is inherently governmental. If these responses are inadequate, interviews, workplace observations, and narrative descriptions of the position can be helpful.

Analyzing text in a functional description consistently and in accordance with the identified rubric is a highly subjective process. To assist analysis, we identified certain language in the functional description that might trigger an IG assessment. Examples of such language are

- "direct/directing teams"
- "manage/managing government employees"
- "responsible for programmatic/budget decisions"
- "draft/drafting responses to Congress or government auditors."

Personal Services

Assessing whether a function constitutes a personal service is more complex than determining what constitutes an IG function. Identifying personal services relies heavily on deciphering workplace relationships. In these assessments, observing the workplace dynamic between contractors and government employees can be very helpful. The questionnaire asks respondents to answer "yes" or "no" to each of the criteria for personal services presented in the FAR and in the definition of personal-services contracts. Responses to these questions—e.g., "Is the contractor under relatively continuous supervision? Does the relationship between government employee and contractor create an employer-employee relationship?"—enable assessment of the key issues in personal-services contracts.

Figure 4.2 shows the process and criteria for an evaluation of personal services. The first step in this process is to assess the employee/employer relationship and whether the position requires some level of supervision from a government official. If the answer to these questions is "yes," then an assessment of whether the personal service constitutes an advisory and assistance service is conducted. This determination requires considerable judgment, demanding that the assessor determine whether a function requires special knowledge or skills. If it is determined that the function is an advisory and assistance function, an assessment of whether the service is "allowed" should be conducted. Our criteria for CAAS determinations and whether they are allowed are derived directly from the FAR.

Other Criteria

The bulk of our analysis centers on assessments of IG and personal-services prohibitions. Yet, the Deputy Secretary of Defense flowchart also lists exemptions, problems with contract administration, and special considerations as grounds for an in-sourcing assessment.[4] At DoD, these are administrative determinations. We do not address these in specific rubrics or approaches, but our questionnaire requests information that could support such assessments.

For example, the questionnaire asks whether the contracted function should have been exempted from performance by contractors. Exemptions from private-sector performance are described in DoDI 1100.22 (and in Chapter Three of this report). Whether an exemption applies can be determined if the person filling out the questionnaire has the knowledge needed to indicate that the position is exempt. If the individual filling out the form does not have this information, other sources must be leveraged.

Similarly, assessments of problems with contract performance can be derived from responses to questionnaire items. In accordance with requirements set out in the FAR and in 10 U.S.C. 2383(a)(2), the questionnaire asks the following: "Is there sufficient organic government expertise to oversee contractor performance of the contract? Are there sufficient control mechanisms and sufficient numbers of military and civilian employees to ensure that contractors are not performing inherently governmental functions? Is there a sufficient number of

---

[4]   It is not clear which special-consideration assessments would warrant an in-sourcing recommendation.

Figure 4.2
A Rubric and Process for Assessing Personal Services

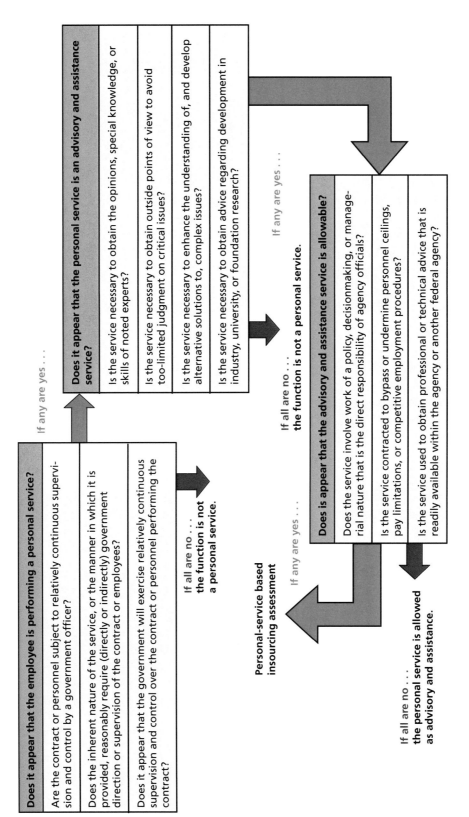

SOURCE: Based on Deputy Secretary of Defense, 2009.
RAND *TR944-4.2*

Contracting Officer Representatives (CORs) appointed to ensure oversight of contract performance?" Again, the usefulness of the information supplied by the respondent depends on his or her level of knowledge.

Current guidance indicates that "special considerations" for certain factors *could* (not must) support in-sourcing decisions. This category includes functions recently performed by federal employees, activities closely associated with IG functions, contracts awarded on a noncompetitive basis, and contracts that have been performed badly. Although the special-consideration provision still exists, we did not assess these criteria.

Limitations and Assumptions

This approach is not part of a total-force assessment. It focuses only on the functions and tasks currently being performed by contractors. It also requires that the current missions, functions, and tasks accurately reflect the goals of the organization and current and expected future work requirements. This approach does not include cost analysis that compares contractor and civilian employees. If a function is judged to be inherently governmental or an unauthorized personal service, cost analysis is not required.

Summary

The analytical approach we propose employs a questionnaire and interviews to collect data relevant to making in-sourcing assessments. Its scope is limited. The legal and regulatory framework, especially the flowchart presented in Deputy Secretary of Defense Lynn's 2009 memorandum, identifies the nature of the function performed and the nature of the workforce relationship as key factors in in-sourcing assessments. Our approach pursues those factors. Our questionnaire allows the analyst to gather information germane to making assessments of IG, CAWIG, and exempt functions. It allows the analyst to ask questions targeting the issue of contractor performance and to collect information about the nature of the contractor–civil-servant relationship in the workplace (as is required during an assessment of personal services). Interviews provide an opportunity to clarify issues arising from the questionnaires and to gather additional information on the nature of the workforce, functions, history, and managerial philosophy of each organization. Observation of the workplace should be part of the assessment process.

CHAPTER FIVE

Conclusions

We looked at the recent history of outsourcing and in-sourcing, assessed the current laws and policies pertaining to in-sourcing, and developed a framework for applying the current laws and policies to arrive at an in-sourcing decision. Our key findings are the following:

- **Historically, sourcing policy has reflected the preferences of the administration at the time.** DoD sourcing's recent history has been dominated by the department's belief that increasing contractor performance of government functions could increase DoD efficiencies and improve mission performance. Clinton- and George W. Bush–era policies expanded opportunities for the private sector in government. Only recently has the emphasis shifted toward the reevaluation of previous outsourcing decisions.

- **In-sourcing and outsourcing have been plagued by the same challenges.** Critics of both have noted the difficulty of developing a strategic and rigorous approach to sourcing decisions; the level of interpretation required to apply existing criteria; and a pervasive lack of data on contracts, numbers of contractors, and functions actually performed. Furthermore, disagreement about how to conduct cost comparisons has led to unresolved controversy. Powerful representatives of private contractors on the one hand and civil-service unions on the other continue to struggle to shape sourcing policy in their constituents' favor, exacerbating the controversy.

- **It appears to be more difficult to in-source than to outsource.** The ease and speed of hiring contractors make outsourcing appealing; in-sourcing efforts (i.e., working within existent civil-service laws to hire contractors as government employees) are considerably more difficult and therefore less appealing. With no clear resolution of these challenges, the future of sourcing policy is like to be as turbulent as its past.

- **In many cases, the laws and policies regarding sourcing are vague, allowing for varied interpretations.** The law (10 U.S.C. 2383) seems to imply a preference for civilian employees over contractors when a function is considered to be CAWIG. Contractors can be used only if there are no government employees available to perform or supervise the work. But, according to USD/P&R policy, CAWIG functions should be assigned to the low-cost provider for purely economic reasons, and contractors may be the low-cost provider regardless of government employee availability. Should a low-cost contractor be chosen even if there is government employee availability? The law is unclear. Multiple definitions of criteria for determining performance of a function increase this problem.

- **There is a gap in the current in-sourcing guidance.** To determine whether a contractor should be in-sourced, we turned to DoD's most recent policy and guidance. We relied heavily on the Deputy Secretary of Defense's May 2009 memorandum, which provided a process and criteria for in-sourcing assessments. However, this guidance fell short of offering those charged with in-sourcing actionable clarification needed to make such determinations. For example, those tasked with in-sourcing are uncertain about how to interpret the IG definition and about how to determine whether a certain function meets this definition.

The analytical approach described in this report is one way to bridge the gap between process and implementation. We have discussed the various incarnations of in-sourcing criteria, and we have offered an interpretation of these criteria. We have also presented a questionnaire that can be used to determine whether certain functions meet various criteria. Although we attempted to create a repeatable analytical process, there is still a great deal of judgment and interpretation required on the part of the in-sourcing analyst. The challenge for policymakers is to develop an in-sourcing methodology that can be used by a wide variety of agencies and that minimizes the opportunity for biased assessments.

Examples of Inherently Governmental Functions, Federal Acquisition Regulation, Subpart 7.5

The following examples of IG functions are reproduced from the FAR, subpart 7.5:

(1) The direct conduct of criminal investigations.

(2) The control of prosecutions and performance of adjudicatory functions other than those relating to arbitration or other methods of alternative dispute resolution.

(3) The command of military forces, especially the leadership of military personnel who are members of the combat, combat support, or combat service support role.

(4) The conduct of foreign relations and the determination of foreign policy.

(5) The determination of agency policy, such as determining the content and application of regulations, among other things.

(6) The determination of Federal program priorities for budget requests.

(7) The direction and control of Federal employees.

(8) The direction and control of intelligence and counter-intelligence operations.

(9) The selection or non-selection of individuals for Federal Government employment, including the interviewing of individuals for employment.

(10) The approval of position descriptions and performance standards for Federal employees.

(11) The determination of what Government property is to be disposed of and on what terms (although an agency may give contractors authority to dispose of property at prices within specified ranges and subject to other reasonable conditions deemed appropriate by the agency).

(12) In Federal procurement activities with respect to prime contracts—

　(i) Determining what supplies or services are to be acquired by the Government (although an agency may give contractors authority to acquire supplies at prices within specified ranges and subject to other reasonable conditions deemed appropriate by the agency);

　(ii) Participating as a voting member on any source selection boards;

　(iii) Approving any contractual documents, to include documents defining requirements, incentive plans, and evaluation criteria;

　(iv) Awarding contracts;

　(v) Administering contracts (including ordering changes in contract performance or contract quantities, taking action based on evaluations of contractor performance, and accepting or rejecting contractor products or services);

　(vi) Terminating contracts;

> (vii) Determining whether contract costs are reasonable, allocable, and allowable; and
>
> (viii) Participating as a voting member on performance evaluation boards.

(13) The approval of agency responses to Freedom of Information Act requests (other than routine responses that, because of statute, regulation, or agency policy, do not require the exercise of judgment in determining whether documents are to be released or withheld), and the approval of agency responses to the administrative appeals of denials of Freedom of Information Act requests.

(14) The conduct of administrative hearings to determine the eligibility of any person for a security clearance, or involving actions that affect matters of personal reputation or eligibility to participate in Government programs.

(15) The approval of Federal licensing actions and inspections.

(16) The determination of budget policy, guidance, and strategy.

(17) The collection, control, and disbursement of fees, royalties, duties, fines, taxes, and other public funds, unless authorized by statute, such as 31 U.S.C. 952 (relating to private collection contractors) and 31 U.S.C. 3718 (relating to private attorney collection services), but not including—

> (i) Collection of fees, fines, penalties, costs, or other charges from visitors to or patrons of mess halls, post or base exchange concessions, national parks, and similar entities or activities, or from other persons, where the amount to be collected is easily calculated or predetermined and the funds collected can be easily controlled using standard case management techniques; and
>
> (ii) Routine voucher and invoice examination.

(18) The control of the treasury accounts.

(19) The administration of public trusts.

(20) The drafting of Congressional testimony, responses to Congressional correspondence, or agency responses to audit reports from the Inspector General, the Government Accountability Office, or other Federal audit entity.

Examples of Inherently Governmental Functions, Federal Acquisition Regulation, Subpart 7.503(D)

The following examples of IG functions are reproduced from the FAR, subpart 7.503(D):

(1) Services that involve or relate to budget preparation, including workload modeling, fact finding, efficiency studies, and should-cost analyses, etc.

(2) Services that involve or relate to reorganization and planning activities.

(3) Services that involve or relate to analyses, feasibility studies, and strategy options to be used by agency personnel in developing policy.

(4) Services that involve or relate to the development of regulations.

(5) Services that involve or relate to the evaluation of another contractor's performance.

(6) Services in support of acquisition planning.

(7) Contractors providing assistance in contract management (such as where the contractor might influence official evaluations of other contractors).

(8) Contractors providing technical evaluation of contract proposals.

(9) Contractors providing assistance in the development of statements of work.

(10) Contractors providing support in preparing responses to Freedom of Information Act requests.

(11) Contractors working in any situation that permits or might permit them to gain access to confidential business information and/or any other sensitive information (other than situations covered by the National Industrial Security Program described in 4.402(b)).

(12) Contractors providing information regarding agency policies or regulations, such as attending conferences on behalf of an agency, conducting community relations campaigns, or conducting agency training courses.

(13) Contractors participating in any situation where it might be assumed that they are agency employees or representatives.

(14) Contractors participating as technical advisors to a source selection board or participating as voting or nonvoting members of a source evaluation board.

(15) Contractors serving as arbitrators or providing alternative methods of dispute resolution.

(16) Contractors constructing buildings or structures intended to be secure from electronic eavesdropping or other penetration by foreign governments.

(17) Contractors providing inspection services.

(18) Contractors providing legal advice and interpretations of regulations and statutes to Government officials.

(19) Contractors providing special non-law enforcement, security activities that do not directly involve criminal investigations, such as prisoner detention or transport and non-military national security details.

Questionnaire

This appendix reproduces the questionnaire RAND developed to assess whether positions meet criteria for in-sourcing.

In-sourcing evaluation form for OPNAV N-8 Staff

On April 6, 2009, Secretary Gates announced that the Department would scale back the role of contractors in support services. On April 8, the Comptroller signed RMD 802, which included the realigning of resources for FY 2010-2014 to decrease funding for contract support and increase funding for approximately 33.4K new civilian manpower authorizations. As a result of such initiatives, the RAND corporation has been hired to aid in the evaluation of in-sourcing for OPNAV. The objective of this inquiry is to gather data to help to support assessments of in-sourcing.

Instructions

Please fill out the table on page 2, "Organizational Information" only once for your organization. However, pages 3-7 pertain to each individual, and this section must be filled out for each individual to be considered for in-sourcing. Duplicate worksheets are not provided for each individual.

Where indicated, place an x in the box to indicate which response is selected.

RAND Contacts

If you have any questions or require further clarification, please contact one of the analysts listed below:

Jessie Riposo	Stephanie Young	Irv Blickstein
RAND Corporation	RAND Corporation	RAND Corporation
1200 South Hayes St.	1200 South Hayes St	1200 South Hayes St
Arlington, VA 22202-5050	Arlington, VA 22202-5050	Arlington, VA 22202-5050
703-413-1100, ext 5162	703-413-1100, ext 5376	703-413-1100, ext 5047
riposo@rand.org	syoung@rand.org	irv@rand.org

Person(s) Completing the Form

Name	Title/Organization	Phone #	Email address

I. ORGANIZATIONAL INFORMATION

Please fill out the following table:

Type of Personnel	Number of Personnel	Contract Manpower Equivalents and Associated Cost
Military Officers – Assigned Elsewhere		NR
Civil Servants- Non OPNAV (reimbursed by OPNAV)		
Civil Servants – Non OPNAV (not reimbursed by OPNAV)		
Command Development Program		NR
Presidential Management Fellows		NR
FFRDC		NR
Other		NR
Contractors		
OPNAV Paid		
Non-OPNAV Paid		
Inter Personnel Government Act (IPA)		
Other		NR

NR= Not Required

Military Officer – Assigned Elsewhere = personnel serving on the N8 staff but permanently assigned to another command.

Non-OPNAV Civil Servants = personnel serving on the N8 staff, but permanently assigned to another command.

For military officers:

      If the officer is assigned elsewhere, please indicate where.

      Please indicate what mission the officer is in support of.

2

II. INDIVIDUAL INFORMATION

1. What is the name of the individual? [_____]

2. What is the individual's function code? [_____] Billet Code? [_____]

3. Is the individual: (please mark an "X" in the appropriate box)
    a. Contractor [_]
    b. Military Personnel [_]
    c. Civil Servant [_]
    d. Other: [_]
        If Other, Please Explain: [_____]

4. Which organization pays the salary and benefits of this individual?
[_____]

5. What is the cost of this individual to OPNAV? [_____]
6. What is the cost of this individual to the Loaning Organization? [_____]

7. Has a cost assessment been performed of this function?   Yes [_] No [_]

8. Please describe the activities and responsibilities of the individual:
[_____]

9. What enduring mission is supported by the function currently being performed by the contractor?
[_____]

10. How long has the function been performed by the individual? (please place an "x" in the appropriate box)

Years					
1	2	3	4	5	>5

	Yes	No
11. Has the function been performed by a civilian anytime in the past decade?		
12. Was the contract awarded competitively?		

13. If the answer to 7 was **YES**, what is the date planned for re-competition? (Please provide response in mm/yy format)

a. Questions Related to Inherently Governmental Functions

	Yes	No
14. Was this position, as defined in the current contract, determined not to be an inherently governmental function?		
15. If **YES** to question 14, has there been a change to functions being performed, policy, legislation, procedure, or other that has led to the possibility of the position becoming inherently governmental?		

16. If **YES** to question 15, please answer the following questions on Inherently Governmental functions.

Does the function include:	YES	NO
Binding the US to take or not to take action by contract, policy, regulation, authorization, order or otherwise		
Determining, protecting, and advancing US economic, political territorial, property, or other interests by military or diplomatic action; contract management or otherwise		
Make decisions significantly affecting the life, liberty, or property interests of private persons		
Commissioning, appointing, directing or controlling officers or employees of the U.S.		
Exerting ultimate control over the acquisition, use, or disposition of the real or personal, tangible or intangible, property of the US, including the collecting, control, or disbursement of appropriated and other federal funds.		

4

b. Questions Related to Prohibited Personal Services

Which, if any, of the following are true of this individual's function:		YES	NO
1	The contractor personnel are subject to relatively continuous supervision and control by a governmental officer.		
2	Contractor is performing on a government site.		
3	Principal tools and equipment are furnished by the government.		
4	Services are applied directly to the integral effort of agencies or an organizational subpart in furtherance of assigned function or mission.		
5	The need for the service provided can reasonably be expected to last beyond one year.		
6	The inherent nature of the service, or the manner in which it is provided, reasonably require (directly or indirectly), Government direction or supervision of contractor employees.		

c. Questions Related to Exemptions from Private Sector Performance

Has the function:		YES	NO
1	Received an exemption from Private Sector Performance		

d. Additional Considerations for In-sourcing

Does the performance involve:		YES	NO
1	Services that involve or relate to budget preparation, including workload modeling, fact finding, efficiency studies, and should-cost analyses, etc.		
2	Services that involve or relate to reorganization and planning activities.		
3	Services that involve or relate to analyses, feasibility studies, and strategy options to be used by agency personnel in developing policy.		
4	Services that involve or relate to the development of regulations.		
5	Services that involve or relate to the evaluation of another contractor's performance.		
6	Services in support of acquisition planning.		
7	Contractors providing assistance in contract management (such as where the		

5

	contractor might influence official evaluations of other contractors).		
8	Contractors providing technical evaluation of contract proposals.		
9	Contractors providing assistance in the development of statements of work.		
10	Contractors providing support in preparing responses to Freedom of Information Act requests.		
11	Contractors working in any situation that permits or might permit them to gain access to confidential business information and/or any other sensitive information		
12	Contractors providing information regarding agency policies or regulations, such as attending conferences on behalf of an agency, conducting community relations campaigns, or conducting agency training courses.		
13	Contractors participating in any situation where it might be assumed that they are agency employees or representatives.		
14	Contractors participating as technical advisors to a source selection board or participating as voting or nonvoting members of a source evaluation board.		
15	Contractors providing inspection services.		
16	Contractors providing special non-law enforcement, security activities that do not directly involve criminal investigations, such as prisoner detention or transport and non-military national security details. However, the direction and control of confinement facilities in areas of operations is inherently governmental.		
17	Private security contractor in operational environment oversees.		
18	Contract interrogators.		
19	Contract logistics support required for weapon systems that deploy with operational units.		
20	Is there sufficient organic government expertise to oversee contractor performance of the contract?		
21	Are there sufficient control mechanisms and sufficient numbers of military and civilian employees to ensure that contractors are not performing inherently governmental functions?		
22	Is there a sufficient number of Contracting Officers Representatives (CORs) appointed to ensure oversight of contract performance?		
23	Are comparable services, meeting comparable needs, performed in your or other agencies (to your knowledge) using civil service employees?		
24	Does this person support a critical assignment, for which there is no civilian readily		

	available?			
25	Does this person help the Department of Defense to meet recruitment and retention objectives which would not be met if converted to civil servant?			
26	Does this person ensure the existence of a ready and controlled source of technical competence?			
27	Does this person participate in a defined-term rotational assignment from another government organization for the primary purpose of professional development?			
28	Does this person help the Department to maintain some core competency which would otherwise be lost?			
29	Would this position be filled within the required timeframe if converted to a civilian position?			
30	If this position was in-sourced, would this position require special monetary or other retention activities which the Department cannot support?			
31	Does this person represents an entity of interest to OPNAV, but is not appropriate on the OPNAV staff payroll?			
32	Is this a temporary function?			

33   Why is this position critical to the success of the OPNAV organization?

34   Why must this position by maintained by the sponsoring/loaning organization?

35   Do you assess this function to be necessary in the Pentagon (please check the appropriate box),

Full	
Half	
Or Quarter Time	

7

Bibliography

Acquisition Advisory Panel, *Report of the Acquisition Advisory Panel to the Office of Federal Procurement Policy and the United States Congress*, 2007.

Bennet, John T., "Panel: DoD Should Cut 111,000 DoD Civilian Jobs," *Federal Times*, July 26, 2010.

Brodsky, Robert, "Pentagon Abandons In-Sourcing Effort," GovernmentExecutive.com, August 10, 2010a. As of November 5, 2010:
http://www.govexec.com/dailyfed/0810/081010rb1.htm

———, "Defense In-Sourcing to Continue at Military Services," GovernmentExecutive.com, September 7, 2010b. As of November 5, 2010:
http://www.govexec.com/dailyfed/0910/090710rb1.htm

———, "Union Blasts Scaled Down Defense In-Sourcing Plan," September 30, 2010c. As of November 4, 2010:
http://www.govexec.com/dailyfed/0910/093010rb1.htm

Carlstrom, Greg, "Tables Turned: Contractors Complain In-Sourcing Tactics Unfair," FederalTimes.com, last updated March 21, 2010. As of November 5, 2010:
http://www.federaltimes.com/article/20100321/ACQUISITION03/3210306/-1/RSS

Congressional Budget Office, *Contractors' Support of U.S. Operations in Iraq*, Washington, D.C., 2008.

Corrin, Amber, "DoD Gets Ball Rolling on In-Sourcing," *Defense Systems*, 2010.

Defense Science Board Task Force, *Human Resources Strategy*, Washington, D.C.: Department of Defense, 2000.

Deputy Secretary of Defense, "Implementation of Section 324 of the National Defense Authorization Act for Fiscal Year 2008 (FY 2008 NDAA)—Guidelines and Procedures on In-Sourcing New and Contracted Out Functions," memorandum, April 4, 2008.

———, "In-Sourcing Contracted Services—Implementation Guidance," memorandum, May 28, 2009.

Gates, Susan M., and Albert A. Robbert, *Personnel Savings in Competitively Sourced DoD Activities: Are They Real? Will They Last?*, Santa Monica, Calif.: RAND Corporation, MR-1117-OSD, 2000. As of May 16, 2011:
http://www.rand.org/pubs/monograph_reports/MR1117.html

Gates, Robert M., "Statement on the Budget to the Senate Armed Services Committee," Washington, D.C., February 2, 2010.

Ginsberg, Wendy, *Pay-for-Performance: The National Security Personnel System*, Washington, D.C.: Congressional Research Service, 2008.

Gore, Al, *From Red Tape to Results: Creating a Government That Works Better & Costs Less*, Washington, D.C.: Office of the Vice President, 1993.

Grasso, Valerie Bailey, *Defense Outsourcing: The OMB Circular A-76 Policy*, Washington, D.C.: Congressional Research Service, 2005.

Hearing Before the Subcommittee on Government Management, Information, and Technology of the Committee on Government Reform and Oversight, House of Representatives, One Hundred Fifth Congress, First Session, on H.R. 719, Washington, D.C.: U.S. Government Printing Office, 1998.

Kesner, Kenneth, "Defense Secretary Says In-Sourcing Hasn't Cut Costs as Hoped; Future of Initiative Uncertain," *The Huntsville Times*, August 29, 2010.

Kettl, Donald F., *Reinventing Government: A Fifth-Year Report Card*, Washington, D.C.: The Brookings Institution, 1998.

Korroch, R. E., *Rethinking Government Contracts for Personal Services*, thesis, Washington, D.C.: George Washington University, 1997.

Luckey, John R., Valerie Bailey Grasso, and Kate M. Manuel, *Inherently Governmental Functions and Department of Defense Operations: Background, Issues, and Options for Congress*, Washington, D.C.: Congressional Research Service, 2009.

Needham, John K., *Sourcing Policy: Initial Agency Efforts to Balance the Government to Contractor Mix in the Multisector Workforce: Testimony Before the Subcommittee on Oversight of Government Management, the Federal Workforce, and the District of Columbia, Committee on Homeland Security and Governmental Affairs, U.S. Senate*, Washington, D.C.: Government Accountability Office, GAO-10-744T, May 10, 2010.

O'Keefe, Ed, "At Homeland Security, Contractors Outnumber Federal Workers," *Washington Post*, February 25, 2010.

Office of Management and Budget, *Inherently Governmental Functions*, Policy Letter 92-1, 1992.

———, Circular No. A-76 Revised 1999, August 4, 1983.

———, "Performance of Commercial Activities," *Federal Register*, Vol. 67, No. 223, November 19, 2002.

———, Circular No. A-76 Revised, May 29, 2003.

Office of Management and Budget, Office of Federal Procurement Policy, "Work Reserved for Performance by Federal Government Employees," web page, undated. As of November 4, 2010: http://www.whitehouse.gov/omb/procurement_work_performance/

Office of the Assistant Secretary of Defense (Public Affairs), "Sec. Gates Announces Efficiencies Initiatives," News Release No. 706-10, August 9, 2010.

Office of the Secretary of Defense, "Estimating and Comparing the Full Costs of Civilian and Military Manpower and Contract Support," Directive-Type Memorandum 09-007, January 29, 2010.

Orszag, Peter R., "Managing the Multi-Sector Workforce," memorandum, July 29, 2009.

Public Law 101-165, Department of Defense Appropriations Act, 1990, November 21, 1989.

Public Law 105-270, Federal Activities Inventory Reform Act of 1998, October 19, 1998.

Public Law 108-375, Ronald W. Reagan National Defense Authorization Act for Fiscal Year 2005, October 28, 2004.

Public Law 109-163, National Defense Authorization Act for Fiscal Year 2006, January 6, 2006.

Public Law 110-28, U.S. Troop Readiness, Veterans' Care, Katrina Recovery, and Iraq Accountability Appropriations Act, 2007, May 25, 2007.

Public Law 110-181, National Defense Authorization Act for Fiscal Year 2008, January 28, 2008.

Public Law 110-417, Duncan Hunter National Defense Authorization Act for Fiscal Year 2009, October 14, 2008 (as amended by Public Law 111-84, National Defense Authorization Act for Fiscal Year 2010, October 28, 2009).

Public Law 111-117, Consolidated Appropriations Act, 2010, December 16, 2009.

Public Law 111-383, Ike Skelton National Defense Authorization Act for Fiscal Year 2011, January 7, 2011.

Quarterman, Cynthia, "Creating a Government That Works Better and Costs Less: How Far Have We Come?" *Mineral Management Service Today*, Vol. 6, No. 2, 1996.

Rostker, Bernard D., *A Call to Revitalize the Engines of Government*, Santa Monica, Calif.: RAND Corporation, OP-240-OSD, 2008. As of May 19, 2011: http://www.rand.org/pubs/occasional_papers/OP240.html

Schwartz, Moshe, *Department of Defense Contractors in Iraq and Afghanistan: Background and Analysis*, Washington, D.C.: Congressional Research Service, 2010.

Simon, Jacqueline, "Statement Before the House Budget Committee on the Department of Defense Efficiency Initiative," September 30, 2010.

Soloway, Stan, "Defense Department's Approach to In-Sourcing Has Unintended Consequences," *Washington Post*, July 19, 2010.

Spoth, Tom, "First Attempts at In-Sourcing Show Challenges Ahead," FederalTimes.com, May 21, 2010. As of November 5, 2010:
http://www.federaltimes.com/article/20100521/ACQUISITION02/5210302/1001

Under Secretary of Defense, "Implementation of Section 343 of the 2006 National Defense Authorization Act," memorandum, Washington, D.C., July 27, 2007.

———, "Better Buying Power: Mandate for Restoring Affordability and Productivity in Defense Spending," memorandum, Washington, D.C., June 28, 2010.

U.S. Department of Defense, "Defense Budget Recommendation Statement," as prepared for delivery by Secretary of Defense Robert M. Gates, Arlington, Va., April 6, 2009.

———, *Quadrennial Defense Review Report*, Washington, D.C., February 2010a.

———, *Policy and Procedures for Determining Workforce Mix*, DoDI 1100.22, April 12, 2010b.

U.S. General Accounting Office, *Defense Advisory and Assistance Service Contracts*, Washington, D.C., B-276026, June 13, 1997.

———, *DoD Personnel: DoD Actions Needed to Strengthen Civilian Human Capital Strategic Planning and Integration with Military Personnel and Sourcing Decisions*, Washington, D.C., GAO-03-475, 2003.

U.S. Government Accountability Office, *Department of Homeland Security: Improved Assessment and Oversight Needed to Manage Risk of Contracting Selected Services*, Washington, D.C., GAO-07-990, 2007.

———, *Defense Management: DoD Needs to Reexamine Its Extensive Reliance on Contractors and Continue to Improve Management and Oversight*, Washington, D.C., GAO-08-572T, 2008.

———, *Civilian Agencies' Development and Implementation of In-Sourcing Guidelines*, Washington, D.C., GAO-10-58R, 2009.

U.S. Office of Personnel Management, "Direct-Hire Authority (DHA) Fact Sheet," web page, undated. As of November 5, 2010:
http://www.opm.gov/directhire/factsheet.asp

U.S. Senate Committee on Homeland Security and Governmental Affairs, *Balancing Act: Efforts to Right-Size the Federal Employee-to-Contractor Mix*, Washington, D.C., 2010.

Weigelt, Matthew, "Obama Hits Campaign Trail to Sell In Sourcing," WashingtonTechnology.com, January 12, 2010a. As of November 5, 2010:
http://washingtontechnology.com/Articles/2010/01/11/COVER-STORY-Insourcing-main-bar.aspx

———, "OFPP Proposes Tests for Deciding When to Outsource Work," WashingtonTechnology.com, March 31, 2010b. As of November 5, 2010:
http://washingtontechnology.com/articles/2010/03/31/ofpp-inherently-governmental-function-fair-act.aspx

———, "Small Business Fights In-Sourcing . . . and Wins," WashingtonTechnology.com, May 5, 2010c. As of May 10, 2011:
http://washingtontechnology.com/articles/2010/05/03/procurement-insourcing-boone-v-air-force.aspx

The White House, "Memorandum for the Heads of Executive Departments and Agencies, Subject: Government Contracting," press release, March 4, 2009.

Wittman, Robert J., Jim Moran, Jeff Miller, Todd Tiahrt, Joe Wilson, J. Randy Forbes, Michael R. Turner, Paul C. Broun, Doug Lamborn, Duncan Hunter, and Bill Posey, letter to Robert M. Gates, Washington, D.C., July 31, 2009.